BOTANICAL JOURNAL OF SCOTLAND

(formerly Transactions of the Botanical Society of Edinburgh)

Vol. 46 Part 4 1994

PLANTS AND PEOPLE

Economic Botany in Northern Europe AD 800-1800

Special Issue of Papers from the Botanical Society of Scotland Symposium held at the Royal Botanic Garden, Edinburgh, 24-27 September, 1993

Edited by
J.H. Dickson & R.R. Mill

Published by
EDINBURGH UNIVERSITY PRESS LTD

B·S·S

Cover design by Paul Nesbitt from John Parkinson's '*Paradisus Terrestris*' (1629)

ISBN 0 7486 0526 6

ISSN 0374-6607

Published by Edinburgh University Press Ltd.
22 George Square, Edinburgh EH8 9LF

Typeset by PVA Department, The Scottish Agricultural College,
West Mains Road, Edinburgh EH9 3JG

Printed by J. W. Arrowsmith Ltd, Bristol BS3 2NT

Editorial Note

Beginning in 1994 the Botanical Journal of Scotland will be published by Edinburgh University Press Ltd. This development has the support of a substantial majority of members of the Society who have expressed a view. It is seen by the Editorial Board as advantageous with respect to the management of production and circulation but the change does not affect editorial policy relating to content of the Journal. With the new publishing arrangements two issues will be produced each year, in June and December. It is hoped that this will be welcomed by readers and contributors, and will be a move towards a reduced time between papers being accepted and published while providing more scope for publishing, as special issues, papers from symposia which are periodically held by the Society.

The present number represents such an issue and is made up of papers presented at the Symposium, 'Plants and People: Economic Botany in Northern Europe AD 800-1800', held in Edinburgh, September, 1993. It may be seen as a departure from normal Journal policy in being concerned with more than the scientific study of plants, but with the much broader context of cultural, economic and social aspects, accommodating a range of disciplines, orientations and perceptions in approaches towards the study of plants. In focusing on the past, much of the subject matter is drawn from archaeological recording, historical research, archive material and oral tradition. The Society acknowledges with gratitude the organizers of a successful conference, Dr. J. H. Dickson and Dr. R. R. Mill, who brought together scholarship from many branches of learning and gave members an opportunity to gain new perspectives on plants and people from an earlier period. This publication, in adopting a less formalized approach than is customary for the Journal, attempts to convey the character of the Symposium, and it is hoped that the issue will serve to bring the essentials of the proceedings to a wider audience.

Preface to 'Plants and People' Symposium Issue

One hundred and sixty-five people attended the symposium which took place from 24-27 September, 1993 in the Royal Botanic Garden, Edinburgh. The main participants reading papers or presenting posters came from Sweden, Poland, Norway, the Netherlands, Ireland, Finland, the Faroe Islands, Denmark and Wales, Scotland and England. This mixture of nationalities gave substantial and satisfying justification to the geographical scope of the symposium. Addressing the meeting, sitting in the audience and joining the excursion were archaeologists, botanists, conservationists, folklorists, foresters, gardeners, herbalists and historians; this made a fruitful, lively and perhaps rare mixture of disciplines.

After the welcome from Miss Jacqui Muscott, President of the Botanical Society of Scotland, the first afternoon of the symposium was devoted to eight papers on archaeobotany, a subject which has greatly flourished in the last twenty years and more. Botanical aspects of the economies of Norse peoples in Scandinavia and Britain were a prominent feature. In the evening after the official welcome to the Royal Botanic Garden by the Regius Keeper Professor D. Ingram, Professor K. Faegri, Honorary Fellow of the Botanical Society of Scotland, delivered his very well received special lecture on the rise and fall of garden plants.

The first morning began with Dr. John Harvey's paper on mediaeval gardening; long familiar with the much quoted book Mediaeval Gardens many in the audience were very pleased to be hearing Dr Harvey for the first time. The morning continued with papers on early botanical collections. Chaired by Dr. P. Macpherson, recent past president of the Botanical Society of the British Isles, the second afternoon dealt mainly with medicinal plants. The second morning, chaired by Dr. C. Easton of Scottish Natural Heritage, began with Dr. O. Rackham's paper on woodmanship; the influence that Dr. Rackham's books on the historical ecology of woodlands has had showed clearly in other papers in the session. Chaired by Professor C. Smout of St. Andrews University, the final session on botanical exploration was ended by Professor William Stearn's paper on Linnaeus. In their liveliness and great knowledge Professors Faegri and Stearn, both octogenarians, were an inspiration to all the participants not least those sixty years younger, men and women in their early twenties. Seventeen posters were an informative addition to the 32 papers which were delivered.

During my presidency of the Botanical Society of Scotland 1990-1992 it was a privilege to have been the instigator of the symposium. Without the ideas and energy of Dr. R. R. Mill, the co-editor of this volume, and of Mr. R. Galt, secretary of the Botanical Society of Scotland, the meeting would have had neither its final, full character nor its smooth running. For the latter I am also much indebted to Ms. Sharon Kirk (Events Organizer at the Royal Botanic Garden), to Mrs. Jane Galt, Mr. W. Marshall and Mrs. Eileen Marshall. I am also very grateful to the Botanical Society of the British Isles for subsidizing Professor Faegri's visit, to Drs. Maria and David Chamberlain and to Mrs. S. Wawro for hospitality to foreign visitors, to the six chairpersons, to His Grace the Duke of Buccleuch for permission to visit Dalkeith, Mr. A. W. Wiseman for permission to visit Cadzow, to the Regius Keeper of the Royal Botanic Garden for use of the facilities of the Garden, and to Mr. P. Nesbitt for designing the publicity material. For leading the excursion to Dalkeith I thank Dr. O. Rackham and Dr. J. Cousens and also Mr. P. Quelch of the Forestry Authority and Mr. S. Hunter for providing the display of wood working and cutting at lunch-time at Cadzow. Finally all the helpful staff, fine facilities and splendid setting of the Royal Botanic Garden were much appreciated by all the participants, as has been the case over the years at previous weekend symposia organized by the Botanical Society of Scotland then called the Botanical Society of Edinburgh.

June, 1994

J. H. Dickson

CONTENTS

The Usage of Wild Berries and Other Fruits in the Mediaeval and Post-mediaeval Households in Norway

KERSTIN GRIFFIN

Arkeologisk museum i Stavanger, PB478, N4001, Stavanger, Norway

Summary

This paper is a discussion of the significance of the recovery of remains of berries and other fruits from mediaeval and later layers at six Norwegian sites including Oslo, Bergen and Trondheim.

There is evidence of the extensive usage of wild berries in Norwegian mediaeval and post-mediaeval households from several archaeological excavations in towns and in the countryside (Fig. 1). The information comes mostly from analyses of latrine deposits and refuse heaps. The present distribution of many species of berries is reflected in the mediaeval material. Fruit stones of sloe (*Prunus spinosa* L.), which today has a south-eastern distribution in Norway, are common in Oslo, Tønsberg and Skien. Cloudberry (*Rubus chamaemorus* L.), is present in the whole country, but most common in northern Norway. Fruit stones of cloudberry dominate the finds from Måsøy, north Norway, they are commonly present in Trondheim and occur in smaller amounts in Oslo and Tønsberg.

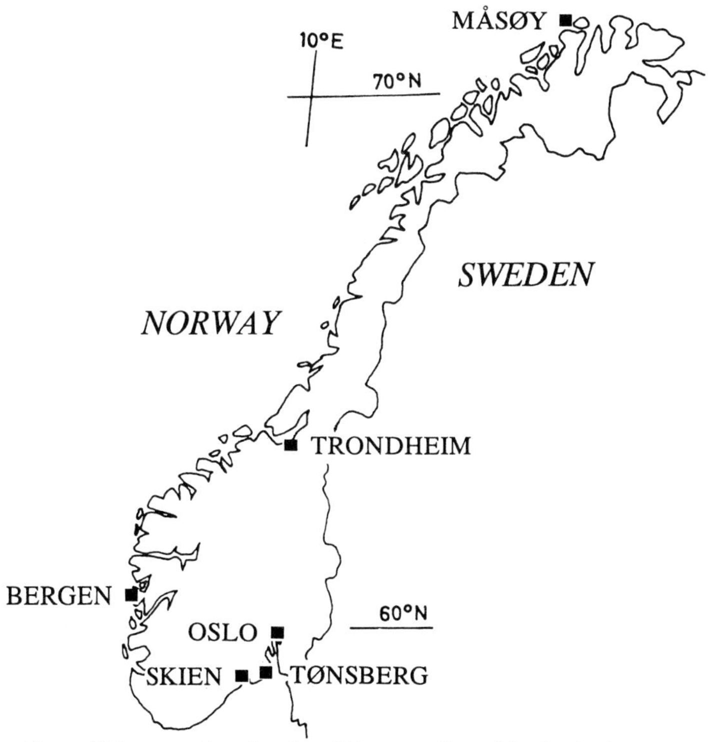

Fig. 1. Map of Norway showing localities mentioned in the text.

There is little evidence for long distance trade. Berries were most likely offered for sale in the local market. Longer transport of berries was probably problematic. In a country like Norway with high mountains and deep fiords and a very long coastline the usage of local natural resources has always been and still is important. The wild berries are one of these resources.

In spite of industrial development and increased urbanization there is plenty of undisturbed land in Norway. Wild berries occur in abundance in the woods, in meadows and on the heaths along the coasts and in the mountains. It is a strong tradition to go out and pick berries in the seasons and to prepare jam, juice and other derivatives from them. In Norway and also in Sweden and Finland everybody has the right to pick berries anywhere (except for cloudberries in north Norway).

The most popular berries are the blueberries (bilberries, *Vaccinium myrtillus* L.) that are common in the whole country. In very good years a few days of picking will yield enough berries to last a family of four for the whole year. Other popular berries are red whortleberries (*Vaccinium vitis-idaea* L.), which are mostly used to prepare a jam used with meat, and cloudberries (*Rubus chamaemorus*).

Cloudberries are an important source of income for inhabitants in the north of Norway where the picking of cloudberries is protected by law in certain areas. No unripe berries can be picked and it is not permitted to pick berries on other people's land.

Wild raspberries (*Rubus idaeus* L.) are also picked, especially along roads in the countryside and in the forests where they grow in abundance in logged areas. In the city of Oslo the health authorities have published a series of maps over the city including large forested areas within the city limits. The vegetation types are described and areas with good berry forests are especially marked to aid the inhabitants in finding the berries. Good areas for mushroom collecting are also described and marked on the maps (Kummen & Larsson, 1981; 1982). When areas were sprayed with chemicals warning signs were posted in the woods. In recent years this spraying has been abandoned. In a new plan for the use of the woods in Oslo the possibilities of recreation are strongly emphasized, and one of the activities is berry picking (Oslo kommun, 1991).

The tradition of using wild berries as part of the diet is still very much alive in Norway. Many of the mentioned wild berries can be bought in the markets and the products are available in the stores. However, the industrial production of jam is often based on imported berries from for instance Poland. This is mostly due to the high wages in Norway. In 1904 berries were picked in Telemark, Norway, for export to England. This has been revived in a new trial project.

It is generally assumed that people used wild berries before there were cultivated ones. The upswing of interest in gardening in the 17th century in Norway (Skard, 1963) may have had a negative influence on the picking of wild berries. This influence is, however, not documented in any way.

In a study by Grøn (1927) summarizing the state of nutrition in Norway before AD 1500 it is stated that the Norwegian farming population had a distaste for wild berries, a tradition carried on from mediaeval and earlier times. However, Holmboe (1929) had a very different opinion. The argumentation goes back to a reference to an Icelandic law (Jonsboks landleiebolk) which forbade people to pick berries on other men's land. One could pick as much as one could eat and carry in one's hands. Similar mediaeval laws in Norway include hazelnuts (*Corylus avellana* L.), hops (*Humulus lupulus* L.) and apples (*Malus sylvestris* (L.) Mill.), but not berries. Grøn concluded that Norwegians did not use wild berries while Holmboe believed that there was no need to restrict the picking of wild berries, because they were so abundant. The plants protected in the laws may have been trading items.

The usage of wild berries has obviously been affected by social class. Ove Arbo Høeg collected information on traditions connected with plants in Norway. From his large collection (Høeg, 1974) the following translated statements (p. 66) were taken 'Wild berries played a small role in the household. One gets on the whole the impression that berries were not highly regarded, and they were not considered food' and 'In the old days blueberries were not used very much in the household, but it did occur'. The people giving the information to Høeg all lived in the 19th century. The traditions they tell of are probably handed down orally through generations, but can also be found in the literature. The older written information was most probably produced by men. Tasks, like berry picking, would be performed by women and children and may therefore not have been considered important and not commented upon (Griffin, 1988). I believe that most of Høeg's informers were men and they did not pick wild berries. Not because they did not like to, but because they did not have the time.

The question whether Norwegians ate wild berries or not in mediaeval time is easily answered when studying mediaeval latrine deposits from Norwegian towns. There is an abundance of fruits and seeds of wild berries in these deposits. It is, however, more difficult to estimate the quantity of wild berries they consumed. How can the finding of a few seeds be interpreted? These are questions archaeologists would like answered. However, there is a great variety in the production of seeds and fruits from different plants. There may be up to 80 achenes on one tiny strawberry (*Fragaria vesca* L.). Thus we must find many thousand achenes to suggest a large intake of strawberries. Raspberries (*Rubus idaeus*) may also have a lot of seeds in each berry. Raspberries grow in nitrate rich environments as we know from the outdoor privy and the garbage heaps. The stems bend down when autumn is approaching and the seeds are close to the soil. In good years the number of berries can be great. Concentrations of raspberry seeds can therefore not immediately be interpreted as remains of human food. The presence of the raspberry bush, however, often indicates disturbed ground which may have been caused by *e.g.* windfall, fire or people.

In Table 1 the presence of wild berries and other fruits in some mediaeval and later deposits in Norway are listed. The analysed samples from Gamlebyen, Oslo (Griffin, 1988), include very well preserved latrine deposits where the whole berries are preserved (Fig. 2).

From the mediaeval sites in Tønsberg and Skien no defined latrine deposits were analysed. The samples came from organic layers which contained waste material which most probably also included latrine deposits but also animal manure.

At one excavation site in Tønsberg *c.* 150 samples were sieved in the field in a coarse mesh (*c.* 2 mm) to get osteological material. The sieving also produced a large number of nutshells, fruitstones and seeds (Griffin, unpublished)

From several mediaeval sites in Trondheim well defined latrines, which have been stationary through several archaeological phases have been sampled and analysed. Large concentrations of achenes of *Rosa* sp. and fruit stones of *Empetrum* sp. were found suggesting *e.g.* production of a beverage (Griffin & Sandvik, 1989).

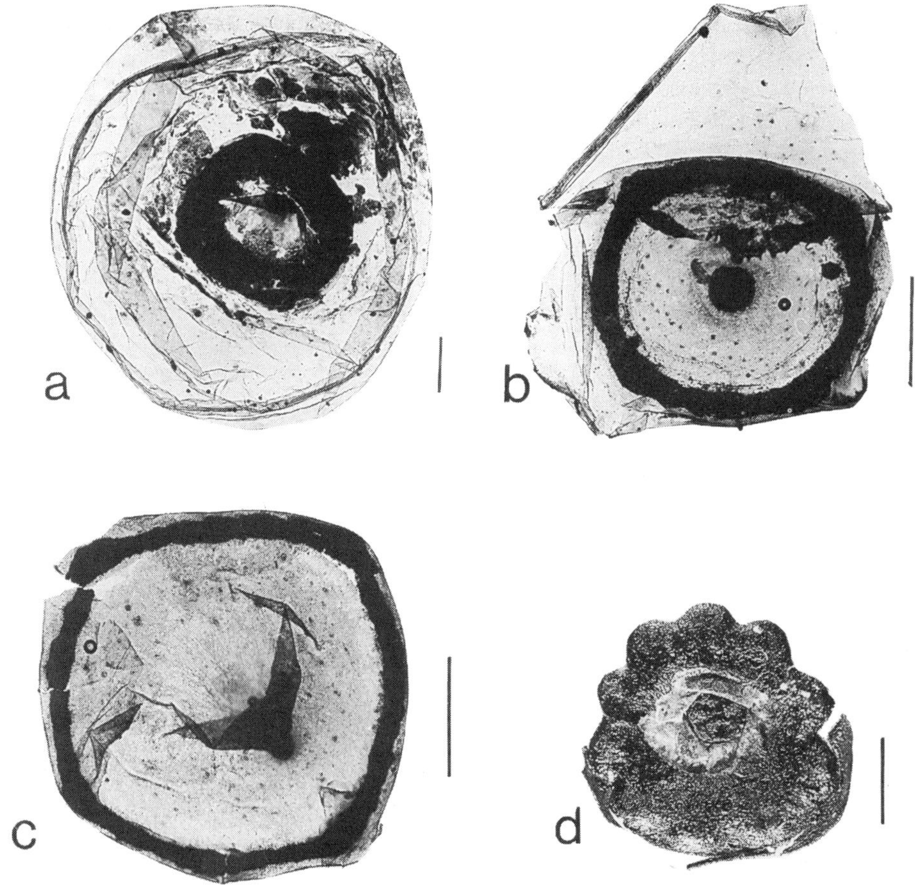

Fig. 2. Remains of berries from latrine deposits in Gamlebyen, Oslo. a-b) cf. *Empetrum* sp., a) berry with fruit wall preserved, b) 'top' of berry with fragments of the fruit wall. c) *Vaccinium* cf. *uliginosum*, 'top' of berry, d) *Vaccinium myrtillus*, 'top' of berry. (Each scale line equals 1 mm. a-c, in glycerine, d, dry). (from Griffin, 1988)

A well defined latrine was excavated in a mediaeval excavation in Bergen. The deposit was stratified and the abundance of seeds varied in the different strata, which could be a result of seasonal variations (Krzywinsky, 1979; Krzywinsky, Fjelldal & Soltvedt, 1983).

At the northernmost site, Måsøy, a farm mound dated from *c.* AD 1600 was excavated and the organic layers analysed for plant remains. The oldest layer which rested on beach pebbles contained cereal bran, fragments of the resistant seed wall of *Agrostemma githago* L., *Rubus chamaemorus*, *Vaccinium myrtillus* and fruit stones of *Empetrum* sp. The combination is interpreted as remains of human food (Griffin, 1985).

Table 1 Wild berries and other fruits found in mediaeval and later deposits in Norway
1=Tønsberg (unpublished reports; Griffin, 1984; Griffin & Foldøy 1986; Foldøy & Griffin, 1986)
2=Skien (Griffin, unpublished, Myrvoll, 1992 (description of site))
3=Oslo (AD 1000-1400: Griffin, 1988; AD 1625-1680: Griffin, 1981)
4=Trondheim (Tallantire, 1979; Griffin & Sandvik, 1989; Griffin, unpublished).
5=Bergen (Krzywinzki, 1979; Krzywinski *et al.*, 1983)
6=Måsøy (Griffin, 1985).

(plant names follow Lid 1985)		1	2	3		4	5		6
	Locality	**Tønsberg**	**Skien**	**Oslo**		**Trondheim**	**Bergen**		**Måsøy**
	Timespan A.D.	800-1300	900 - late 13th C.	1000-1400	1625-1680	1000-1300	14th C.	1600	1600-1944
Arctostaphylos uva-ursi (L.) Spreng.		-	-	+	-	+	+	+	+
Cornus suecica L.		+	-	-	-	+	+	+	+
Corylus avellana L.		+	+	+	+	+	+	+	-
Crataegus monogyna Jacq.		+	-	+	-	-	-	-	-
Empetrum sp.		+	+	+	-	+	+	+	+
Fragaria vesca L.		+	+	+	+	+	+	+	-
Juniperus communis L.		+	-	+	-	+	+	+	+
Malus sylvestris Mill.		+	-	+	+	+	+	+	-
Oxycoccus quadripetalus Br.-Bl.		+	-	+	-	+	-	-	-
Prunus padus L.		+	-	+	-	+	-	+	-
Prunus spinosa L.		+	-	+	-	-	-	-	-
Rosa sp.		+	+	+	-	+	-	-	-
Rubus caesius L.		-	-	-	-	-	+	-	-
Rubus chamaemorus L.		+	-	+	+	+	+	+	+
Rubus idaeus L.		+	+	+	+	+	+	+	-
Rubus cf. plicatus Weihe & Nees		-	-	+	-	-	-	-	-
Rubus cf saxatilis L.		+	-	+	-	-	+	-	-
Rubus sp.		+	-	-	-	-	+	+	-
Sorbus aucuparia L.		+	+	+	-	-	+	+	-
Vaccinium myrtillus L.		+	-	+	-	-	-	-	+
Vaccinium vitis-idaea L.		-	-	+	-	-	-	-	-
Vaccinium uliginosum L		-	-	+	-	-	-	-	-
Vaccinium spp.		+	-	+	+	-	+	+	+
Viburnum opulus L.		+	-	+	-	-	-	-	-

References

Foldøy, O. & Griffin, K. (1986). Paleobotaniske analyser fra Øvre Langgate 65/67, Tønsberg. In *Innberetning over de arkeologiske utgravningene i Øvre Langgate 65/67. Tønsberg.* (Blohme, M. & Runeby, C.) 36 pp. Unpublished report. Bilag 5.

Griffin, K. (1981). Planterester fra Revierstredet 5 og 7. *Riksantikvarens Skrifter* **4,** 272-282.

Griffin, K. (1984). Botanisk prøvematerial. 1. Analyse av 12 jordprøver (transektprøver) fra utgravningene i Storgaten 24/26, Tønsberg 1979. 2. Analyse av 23 jordprøver fra utgravningene i Storgaten 24/26, Tønsberg, 1979. In *Innberetning over de arkeologiske utgravningene i Storgaten 24/26, Tønsberg, 1979.* (Lindh, J), 15 pp. Unpublished report. Bilag 10.

Griffin, K. (1985). Stratigraphic studies in farm mounds in North Norway. *Iskos* **5**, 191-204.

Griffin, K. (1988). Plant remains. *De arkeologiske utgravninger i Gamlebyen, Oslo* **5**, 15-108.

Griffin, K. & Foldøy, O. 1986. Botaniske analyser av jordprøver fra Baglergaten 2/4. In *Innberetning over de arkeologiske utgravningene i Baglergaten 2-4, Tønsberg, 1981-1982.* (Brendalsmo, J), 50 pp. Unpublished report. Bilag 11.

Griffin, K. & Sandvik, P.U. (1989). Frukter, frø og andre makrofossiler. Funksjoner og aktiviteter belyst gjennom analyser av jordprøver, 108 pp. *Fortiden i Trondheim bygrunn: Folkebibliotekstomten, Meddelelser* 19.

Grøn, F. (1927). Om kostholdet i Norge indtil aar 1500. *Det Norske Videnskaps Akademi i Oslo, Skrifter II Hist.-Filos. Klasse 1926,* No. 5 Oslo.

Holmboe, J. (1929). Gamle norske matplanter, 36 pp. *Det Norske Videnskaps-Akademi i Oslo. I. Matem.-Naturvid. Klasse* 1929. No. 2.

Høeg, O.A. (1974). *Planter og tradisjon.* Oslo: Universitetsforlaget.

Krzywinski, K. (1979). Preliminær undersøkelse av planterester i latrine. *Arkeo* 1979, 33-39.

Krzywinski, K., Fjelldal, S. & Soltvedt, E.-C. (1983). Recent palaeoethnobotanical work at the mediaeval excavations at Bryggen, Bergen, Norway. In *Site, environment and economy* (ed. B. Proudfoot), pp. 145-169. *BAR International Series 173.*

Kummen, T. & Larsson, J.Y. (1981, 1982). *Vegetasjonskart over Oslo.* Oslo Helseråd, Kontoret for natur- og miljøvernsaker.

Lid, J. (1985). *Norsk, Svensk, Finsk Flora,* 837 pp. Det norske samlaget. Oslo.

Myrvoll, S. (1992). Handelstorget in Skien - a study of activity on an early medieval site, 311 pp. *Nytt fra Utgravningskontoret i Bergen (N.U.B.),* No. 2.

Oslo kommun (1991). *Mål og retningslinjer for Forvaltning av Oslo kommunes skoger,* 48 pp. Oslo.

Skard, T. (1963). *Hagebruk og gartneri i Norge.* Oslo.

Tallantire, P.A. (1979). Late Viking and early medieval plant material from Trondheim - a problem in interpretation. *Archaeo-Physika* **8**, 295-301.

Erkebispegården, The Archbishop's Palace, Trondheim, Norway, through Late Mediaeval and Post-mediaeval Time

PAULA UTIGARD SANDVIK

Directorate of Cultural Heritage, Trondheim Archaeological Unit, Kjøpmannsgt. 25, N-7012 Trondheim, Norway

Summary

The Archbishop's Palace was an important part of the town of Nidaros (Trondheim) both in mediaeval time and after the Reformation. Considerable knowledge of the buildings and activities has been gained since the excavations started in 1991. Botanical, entomological and osteological analyses have enabled a large, wooden-framed tank to be interpreted as a fish pond.

Background

Parts of the Archbishop's palace in Trondheim were destroyed in a fire in 1983. Before erection of new buildings, archaeological excavations at this site were started 1991 and are planned to be finished by the end of 1994. This presentation concentrates on the preliminary results of the palaeoecological research and concerns only a few aspects of work on this site.

The Archbishop's Palace and its surroundings

The town Trondheim is situated beside the Trondheimsfjord in the country of Sør-Trøndelag (Fig. 1). The landscape has undergone many changes during the last 10,000 years since the Weichselian ice disappeared from the area (Reite, 1983). The shoreline displacement is 180 m (Kjemperud, 1981). Large areas are covered by marine clays and erosion, transport and resedimentation have built up deltaic deposits at the mouths of the rivers.

The town Trondheim dates back to AD 997 and the mediaeval part is situated on the delta of the river Nidelva. The old name of the town, Nidaros, means 'at the mouth of the Nid'. The highest part of Nidarneset, in the south-east is 16 m above sea level. The sandy deltaic deposits on this part are covered by 2-6 m thick clay layers brought to the area as a result of landslides. The oldest known landslide took place between AD 90-410 (Sandvik, 1990) and geotechnical investigations show evidence of three or four altogether (Sand, 1991). According to the seashore displacement curve (Kjemperud, 1981), this particular area has not been dry land more than *c.* 2,500 years. The Cathedral and the Archbishop's Palace were built on top of the consolidated clay. The mediaeval settlement was concentrated on the eastern part of Nidarneset towards the river (Fig. 1).

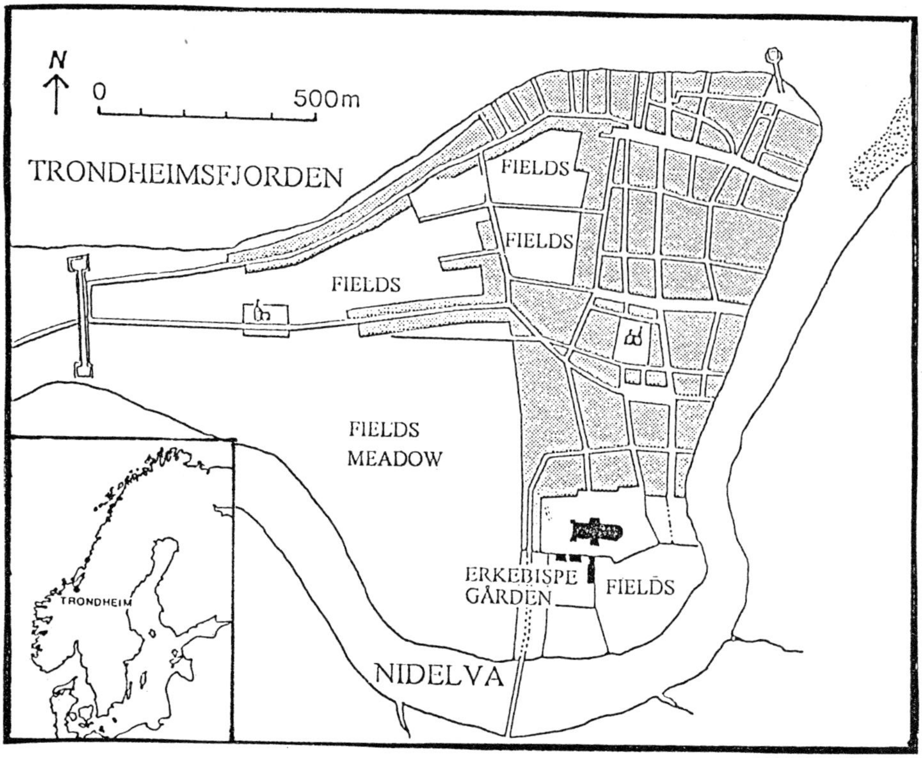

Fig. 1 Geographical location. Trondheim and Nidarneset with the Cathedral, the Archbishop's Palace and the mediaval settlement (Based on map from Long, 1975).

The building of the Archbishop's Palace started in *c.* 1170. After the Reformation in 1537 when the last archbishop Olav Engelbrektson had left Norway, the Palace was the residence for 'Lensherren', the king's representative in this part of Norway. There are very few written descriptions of the palace during high and late mediaeval time, while some of the buildings from the Lensherre period are relatively well known. During the excavation it was very satisfying to find remains which seemed to fit those documented structures. However, the excavation also revealed many unmentioned structures. Most of these were of more or less familiar types; the building details and various finds allowed understanding of their function. There were structures which proved difficult to interpret; one of these is discussed here (Fig. 2).

Fig. 2. The tank with the wooden frame seen from west. The middle part cut by a cellar dated to the 1700s. In the northern part the logged well is seen. Photo: Edwin Baker, Directorate of Cultural Heritage.

'The Carp Pond'

This structure is a big tank, 14 x 4 m with a wooden frame (Fig. 2), filled with sandy-silty clay mixed with different types of organic remains. A vertical cut west to east through the layers in the southern part is shown in Fig. 3. Part of a skeleton of a fish was found in layer D432 and osteological analyses showed that this was a carp. It was impossible to tell whether it was a common carp (*Cyprinus carpio* L.) or a crucian carp (*Carassius carassius* (L.)).

Our question was: have we found the remains of a late mediaeval fish pond?

Unfortunately the wood in the logged frame, which was Scots pine (*Pinus sylvestris* L.), was for various reasons unsuitable for dendrochronological dating (Terje Thun, oral comment). The wood from the logged well (Fig. 2), also Scots pine, is dated to post 1583. According to the stratigraphy (McLees, in prep., Petersèn, in prep.) the construction may be dated back to 1532 which means that it originates from late mediaeval time or early post-Reformation time. The archbishop Olav Engelbrektson's accounts (Olav Engelbriktssons rekneskapsbøker, 1532 to 1538) tell that dried crucian carp was a part of the diet for his men.

Fossils

The analyses not yet being complete, the results presented in Table 1 are preliminary and based on analysis of macrofossils (diaspores and some zoological remains) and microfossils (pollen, spores and diatoms). These fossils are used as indicators, showing differences and similarities between the sediments in the layers.

Macrofossils

Botanical macrofossils from the different layers in the fill were analysed (Fig. 3, Table 1). Standard books and papers were used for making the identifications. The term diaspores is used both for seeds and pips.

The samples from the upper layers were rich in diaspores, mainly from fruits and berries, but also weeds. The samples from the lower layers were relatively poor in diaspores and few food plants were represented. On the basis of identified plant remains there is a clear difference between an upper and a lower unit in the layers (Table 1).

Zoological macrofossils were also recovered during the analyses of botanical remains. As shown in Table 1 they are roughly grouped; complete analyses will be done later by specialists. The only identified type is ephippium from *Daphnia* sp., animals which live their entire life in fresh water (Øakland, 1975). The identification was confirmed by Dr John O. Solem, Museum of Natural History and Archaeology, University of Trondheim. No ephippia were seen in the upper layers, but they were abundant in the lower layers, apart from D433. Insects and fly puparia were common in all layers.

Microfossils

Samples from all layers were prepared for pollen analysis according to the standard procedures. Pollen, spores and charcoal dust were analysed in most of the samples, but these analyses are not finished. Charcoal particles dominated the samples from the upper layers together with different pollen types, cereals (Cerealia), weeds and grasses (Poaceae), but no pollen type was dominating. The lower layers were dominated by pollen from grasses, some cereals and weeds together with lots of charcoal particles.

Diatoms were found abundantly in samples from the lower layers apart from D433, while some few fragments which were impossible to identify were seen in the samples from the upper layers. No marine diatoms were identified.

Discussion

On palaeontological grounds, it is possible to divide the layers into two main units (Fig. 3, Table 1). The upper one is dominated by remains of food plants and charcoal dust. Plants found in the local flora are strawberries (*Fragaria vesca* L.), raspberries (*Rubus idaeus* L.) and cloudberries (*Rubus chamaemorus* L.). From early mediaeval time until now there is evidence for an unbroken tradition for use of wild berries for food in Norway (Griffin, this volume). Results from previous investigations of botanical macro-remains in Trondheim (Tallantire, 1979; Griffin and Sandvik, 1991; Sandvik, 1993, Sandvik in prep., Griffin unpubl.) gives support to this. Nordhagen (1941) tells about the monastery at Tautra, a small island in Trondheimsfjorden in the county of Nord-Trøndelag, which was famous for its fruit gardens where cherries (*Prunus avium* L. and /or *P. cerasus* L.) were grown. The

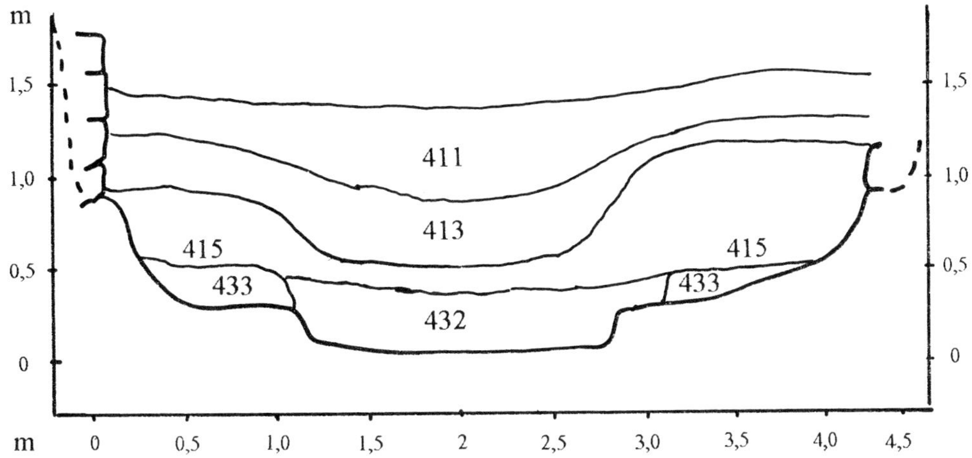

Fig. 3. A west to east cut through the layers in the southern part of the tank.

Table 1. Fossil types identified in the sediments from the tank. XXX = dominating, XX = common, X = present

	MACROFOSSILS					MICROFOSSILS		
Unit/ context	Diaspores from food plants	Diaspores from weeds and others	Insects	Fly puparia	Ephippia	Pollen	Charcoal dust	Diatoms
Upper unit D411, D413	XX *Ficus carica* *Fragaria vesca* *Prunus* cf. *avium* *Rubus chamaemorus* Rubus idaeus	XX *Stellaria media* *Hyoscyamus* cf. *niger* *Papaver somniferum*	XX	XX		XX *Urtica* *Humulus* Cereals Poaceae	XXX	
Lower unit D415, D432, D433	X *Ficus carica* *Fragaria vesca* *Rubus idaeus*	XX *Chenopodium album* *Stellaria media* *Urtica urens*	XX	XX	XX	XX Poaceae Cereals Weeds	XX	XX

cherries thus may have been grown in Trøndelag. Diaspores from figs (*Ficus carica* L.) which are common in these layers show trading contacts with southern Europe. The sweetness of dried figs was much appreciated. These fossils of edible plants came from latrine fills and represent the diet. Henbane (*Hyoscyamus* cf. *niger* L.) and opium poppy (*Papaver somniferum* L.) are among the very rare finds here, while other weeds are common.

Some of the identified fossils originate from organisms, both plants and animals, living in water. The fish skeleton, ephippia and many types of diatoms are located to the layers D415 and D432 which seem to have been formed in water. The skeleton may, of course, be refuse, but may also be remains of a fish which lived and later died in this tank. The bottom layer D433 seems to have been established before the aquatic phase, possibly through the construction of the tank. From the results so far, the best interpretation of the function of this construction is that this was a late mediaeval or early post-Reformation fish pond. To test this conclusion results of analyses of the sediments formed in known fish ponds are required.

References

Griffin, K. and Sandvik, P.U. (1991). Plant remains from mediaeval Trondheim, Norway. *Acta Interdisiplinaria Archaeologica Tomus VII,* 111-115.

Kjemperud, A. (1981). A shoreline displacement investigation from Frosta in Trondheimsfjorden, Nord-Trøndelag, Norway. Norsk Geologisk Tidsskrift **61,** 1-15.

Long, C.D. (1975). Excavations in Trondheim 1971-1974. *Zeitschrift für Archäologi des Mittelalters* **3,**183-207.

McLees, C. (in prep.). Excavations in Erkebispegården 1992. Areas 1C: Stratigraphy and phasing. *Arkeologiske undersøkelser i Trondheim* no. 11.

Olav Engelbriktssons rekneskapsbøker (1532-1538). Utgjeve ved Jens Arup Seip.

Nordhagen, R. (1941). De gamle klosterhager. *Foreningen til Norske Fortidsminnesmerkers Bevaring, Årsberetning 1939,* 55-94.

Petersèn, A. (in. prep.). Utgrävningar i Erkebispegården 1991. Delfält 1D: Stratigrafisk analys og fasinndelning. *Arkeologiske undersøkelser i Trondheim* nr. 10.

Reite, A.J. (1983). Trondheim. Beskrivelse til kvartærgeologisk kart 1621 IV. *Norges geologiske undersøkelse 391.*

Sand, K. *(1991). R.832.* Erkebispegården. Grunnundersøkelser. *Geotekniske vurderinger. Geoteknisk seksjon, Plankontoret, Trondheim kommune.*

Sandvik, P.U. (1990). Ytre Kongsgård TA 1988/10. Paleoøkologiske undersøkingar. *Rapport. Riskantikvaren, Utgravningskontoret for Trondheim.*

Sandvik, P.U. (1993). Utgravingane i Erkebispegården 1991. Delfolt IA og IB: Botaniske analyser. *Arkeologiske underspøkelser i Trondheim* nr. 8.

Sandvik, P.U. (in prep.). Utgravingane i Erkebispegården 1991. Delfelt 1C, 1D og 1E: Botaniske analyser. *Arkeologiske undersøkelser i Trondheim* nr. 12.

Tallantire, P.A. (1979). Late Viking and Early Medieval plant material from Trondheim - a problem in interpretation. *Archaeo-Physika* **8,** 295-300.

Økland, J. (1975). *Ferskvannsøkologi.* Universitetsforlaget, Oslo, Bergen, Tromsø.

Archaeobotanical Evidence for Food Plants in the Poland of the Piasts (10th-13th Centuries AD)

M. POLCYN

Palaeoecological Laboratory of the First Piasts Museum, 62-261 Lednógora, Poland

Summary

The period between the 10th and 13th centuries is denoted in the history of Poland as the latter part of early Middle Ages. Politically, it covers the time span marked by the foundation of the Polish state under the Piast dynasty, on the one hand, and the first attempts to unite the country after its dismemberment, on the other. Socially, the most important landmark of this era is the formation of the early feudal structure. Artistically, it is characterized by the development of the Romanesque style. Archaeological excavations carried out in numerous early mediaeval settlements and strongholds have provided rich archaeobotanical sources which allow one to trace the inter-relations between man and plants. Plant remains help one to determine which species were preferred and cultivated.

As far as geography is concerned, the paper deals with the area situated between the Carpathian mountains and the Baltic sea and between the Oder and the Vistula rivers. Since the 6th century this land had been inhabited by Slavonic peoples sometimes referred to as proto-Polish because it was from them that later Polish tribes emerged. Only the area of contemporary Poland that lies between the Vistula and Pasłęka rivers was ethnically different from the rest, since it was inhabited by West Baltic tribes.

Gallus the Anonymous, an early mediaeval chronicler, wrote that 'The country of the Poles is situated far from trade routes and is hardly known abroad save for the merchants who cross it on their way to Ruś.' These words reveal how Poland was perceived by other nations until the 12th century when the chronicler lived.

Apart from the immediate subject of plants and people, it is necessary to illustrate the social background of the Poles and their country between the 10th and 13th centuries.

The process resulting in the formation of the Polish state was the effect of economic, social, and political changes occurring in these lands from the 6th to the 9th centuries. The basic condition of economic development was the evolution of the Slavs from a nomadic community into a permanently settled one. Initially settlements evolved around strongholds. Open settlements were constructed providing services for the stronghold. The inhabitants worked as craftsmen or tradesmen. These settlements were enclosed by walls and later became towns.

The towns were also surrounded by open settlements of a rural character which provided them with agricultural products. It is generally assumed that tillage zones developed concentrically around the settlements. The intensity of tillage exploitation was inversely proportional to the distance between the settlement and the cultivated land. Orchards and gardens were adjacent to the farmstead. They bordered upon small plots that were not allowed to lie fallow. These plots were manured by penned-up cattle and were followed by larger plots cultivated by means of two-crop rotation. The remotest zone constituted patches of land cleared of forest.

Historians consider that the 12th and 13th centuries in Poland saw an increase in overall agricultural production. It is generally considered that the development of colonization and economic progress were effected by feudal landowners and the introduction of a three-crop rotation. Also, the replacement of the primitive plough by one with a metal mouldboard was of great importance as the spread of the latter allowed for the intensification of tillage and an increase in its area.

The first illustrations showing farm work (people in fields and gardens) have come down to us from the 13th century. In them we can find scenes of picking of fruit including grapes and harvesting using a sickle. The seasonal rhythm of farm work was often presented in almanacs. So far, however, the most significant information about plant cultivation has been provided by archaeobotanical data. The following species list of plants grown in Poland in the early Middle Ages has been compiled according to the botanical material from Gniezno (Jaroń, 1939). Tenth century Gniezno was the centre around which the Polish state was formed. From archaeological excavations in Gniezno several metre-thick settlement layers were uncovered. As regards the plant remains that were found there, Gniezno can be treated as representative of the whole country.

Grain crops

Among the grain sown were rye, (*Secale cereale* L.), wheat (*Triticum*), oats (*Avena*), barley (*Hordeum*), and millet (*Panicum miliaceum* L.). Rye was universally introduced in the Middle Ages. At the end of the 11th century, it became the most popular bread grain. Millet is called 'the Slavonic grain'. It is not a bread grain, but it was the most important ingredient in cooking porridge and gruels. Millet, and particularly its chaff, has been found in many places and in large quantities. In Gniezno millet even formed thin layers consisting of grain, chaff, and bran. Also, small lumps of charred millet have been discovered. Excavations often reveal caryopses of *Hordeum vulgare* L. In Gniezno, archaeobotanists came across hulled grains of barley, the majority of which belong to the six-row form. Like rye, oats became a popular cultivated grain as late as the early Middle Ages. Two species of wheat have been reported in Gniezno: *Triticum vulgare* Vill. and *Triticum compactum* Host. They belong to ones most frequently found in archaeological sites dealing with the early Middle Ages in Poland. In Gniezno both species were represented by two types of grains: small and large.

Pulse crops

Among pulses, peas (*Pisum sativum* L.) are most frequently reported at archaeological sites dealing with the early Middle Ages. In Gniezno peas have also proved to be the most numerous among leguminous plants. Characteristic oval seeds with small diameters have been included among small-seed varieties. Lentils, *Lens culinaris* Medik have been among the less numerous pulse plant remains found in Polish mediaeval archaeological contexts. Seeds of lentils from Gniezno belong to a small-seeded form. Broad beans (*Vicia faba* L.) have been least frequently reported in early mediaeval excavations in Poland. These seeds have not been found in Gniezno.

Vegetables and tubers

The uncharred seeds of cucumbers (*Cucumis sativus* L.) come mostly from organic layers. The size of the seeds revealed in Gniezno corresponds to the smaller seeds of present-day cucumbers. Cucumber seeds have only rarely been found, mainly around large borough excavation sites. They are a feature of an advanced horticultural level. Cucumber growing was probably most popular in what is now the Great Poland region.

Archaeobotanists have rarely found both charred seeds of turnip (*Brassica campestris* L.) at archaeological bog sites or uncharred ones at dry settlement sites. There do arise justified doubts, however, whether the mother plants were cultivated or wild since both species can be found in the wild. There is no doubt that turnip (Polish: '*rzepa*') was well known in that period. A popular Slavonic female name derives from the word denoting turnip: *Rzepicha*. Carrots (*Daucus carota* L.) apart from cultivated forms, may have grown abundantly in the wild. Their characteristic fruit has been found most frequently in the uncharred state in culture layers of archaeological bog sites. Unfortunately, it is not possible to determine, on this basis, which forms they belonged to and whether or not they were grown in early mediaeval times.

Oil and fibre crops

Uncharred flax (*Linum usitatissimum* L.) seeds, pieces of capsules and fibres have very often been found, particularly on waterlogged sites or in organic culture layers. In the early Middle Ages flax was the main fibre plant. It also must have been used as an oil plant and as an ingredient in baking bread. In archaeological sites hemp (*Cannabis sativa* L.) has been found as uncharred achenes. In the early Middle Ages hemp became an important technological plant used in the production of thick cloth. Fragments of such cloth have been discovered in Gniezno. Most probably, also *Papaver somniferum* was cultivated, although only single seeds of opium poppies have infrequently been found on a few archaeological sites.

Fruit trees and nuts

One of the most important plants grown in the early Middle Ages requiring top-level horticultural skills was the grape-vine (*Vitis vinifera* L.). Its cultivation spread along with Christianity. Grape wines were used in church liturgy; as a result, the technology of grape vine cultivation was, most probably, first known to monastic orders for the production of sacramental wine. The other alternative was to import wine from the south of Europe which was particularly difficult at that time. Grape seeds have very rarely been found in Poland (Fig. 1). Nevertheless, historians draw our attention to early mediaeval names of places around Gniezno, Poznań, and other bishoprics and monasteries: *Winiary, Winnica* (Eng. 'vineyard'), *Winnagóra* (Eng. 'vinehill'), Winogrady (Eng. 'grape-vine'). Scenes of grape gathering were depicted on the border of the 12th century Gniezno Cathedral Bronze Door, a unique artifact of that time.

Apple (*Malus*) and pear (*Pyrus*) pips have frequently been found on archaeological sites. Excavations in Gniezno have provided a large and varied collection of pips. Most probably, they came from a number of tree and bush species, either cultivated or growing wild near human settlements. Numerous apple seeds in every culture layer point to the great interest in fruit growing of the settlement's inhabitants. It is very difficult to differentiate particular species on the basis of pips alone since apple culture was rather primitive and produced a multitude of forms characterized by a great variety and mutability of morphological features, including hybrids. Apple trees may have either been wild or cultivated, or transitional ones. Accordingly, it is possible to determine only the genus of the pips.

Fig. 1. Uncarbonized seeds of grape-vine (*Vitis vinifera*) from Gniezno (x 6). Photograph by M. Józwikowska, First Piasts Museum.

As in the case of apples, pear pips have also been found. It is possible to distinguish them from apple seeds thanks to morphological features of their surface. Still, it is difficult to specify whether they are from cultivated or wild plants. Stones of plums (*Prunus domestica* L. subspecies *insititia* (L.) C.K. Schneider) are very common in mediaeval excavations. Stones of plums from Gniezno have been divided into two groups: small stones and typical large ones. The size of both forms of fossil stones is similar to the size of stones of corresponding present-day types. Peach (*Prunus persica* (L.) Batsch.) stones are known from many early mediaeval sites in Poland, particularly those around and in early urban centres. The shape of the stones found in Gniezno corresponds to those of the present day whereas their sizes only partially conform to contemporary ones. Sweet/sour cherry (*Prunus cerasus* L.; *P. avium* L.) stones have often been found in organic culture layers. The variability of the cherry stones' morphological features is so high that it is very hard to determine a species. The finding of walnuts (*Juglans regia* L.) has been recorded on several sites all over Poland. The nuts found on mediaeval sites are considered to have been grown or obtained as the result of trade. Blackthorn (*Prunus spinosa* L.) is a long-standing element of native Polish flora. Judging by the number of fruit stones found, it is possible to conclude that this plant might have been preferred by early mediaeval man because of its edible fruit. Hazel nuts (*Corylus avellana* L.) are a frequent find on many sites. In Gniezno they have been located in all layers. Hazel bushes were a common element of forests where people gathered food. It is not out of the question that they may have been taken into cultivation for their nutritious nuts.

Fruit collected from the wild

Single blackberry (*Rubus fruticosus* L.) seeds have been discovered in Gniezno. Most probably, the fruit was eagerly gathered. Raspberry seeds (Rubus *idaeus* L.) seeds have been found as often as those of blackberries. It is difficult to determine whether raspberry was cultivated or merely gathered. Like hazel nuts, raspberries, blackberries and elderberries (*Sambucus*) were easily accessible. They may have simply fallen on the settlement, or they may have been gathered. In early mediaeval times, tasty fruits of the *Fragaria* species must have been willingly collected from the wild. So far, the greatest amount of wild strawberries have been found in the culture layers of mediaeval Kraków (Wasylikowa, 1978; Wieserowa, 1979) and in waterlogged sites in Great Poland. They have been distinguished as two species: *Fragaria viridis* Duchesne and *Fragaria vesca* L.

The hop plant (*Humulus lupulus* L.) is native to Poland. The presence of its seeds does not necessarily indicate its cultivation. It must, however, have been grown for the production of beer which was widely consumed. This is confirmed by mediaeval illustrations which show barrels of beer together with their connoisseurs.

Now, coming back to the problem of tillage zones around the village, a reconstruction of the crop structure may be attempted. Orchards and gardens adjacent to the farmsteads were planted over with grape vines, and the trees of apples, pears, plums, peaches, sour and sweet cherries, as well as walnuts. Most likely, hawthorn and hazel nuts were also grown. From the pulse crops, peas, lentils, and broad beans were cultivated. Of vegetables, cucumbers, turnips and perhaps, carrots were grown. Small plots may have been planted with crops like flax, hemp or poppy. And, lastly, the most remote plots were used for the cultivation of grain.

Acknowledgements

I would like to thank Professor Krystyna Wasylikowa for sharing her data with me and the Institute of Botany in Kraków for providing me with botanical materials from excavations in Gniezno. I am also indebted to Andrzej Bursa and Jacek Fabiszak from the School of English, A. Mickiwicz University, Poznan, for translation of this text. Many thanks to my wife Ilona for all of her help.

References

Jaroń, B. (1939). Średniowieczne szczątki roślinne z wykopalisk w Gnieźnie. *Gniezno w zaraniu dziejow (VIII-XIIIw) w swietle wykopalisk. Poznan, 273-316.*

Wasylikowa, K. (1978). Plant remains from early and late mediaeval time found on the Wawel Hill in Cracow. *Acta Palaeobotanica* **19**(2), 115-200.

Wieserowa, A. (1979). Plant remains from the early and late Middle Ages found in the settlement layers of the Main Market Square in Cracow. *Acta Palaeobotanica* **20**(2), 137-210.

Some Botanical Aspects of a Norse Economy in Northern Scotland

JACQUELINE P. HUNTLEY

Department of Archaeology, University of Durham, Science Laboratories, South Road, Durham, DH1 3LE

Summary

In northern Scotland, there is a change from naked to hulled barley sometime around the birth of Christ, although the reason is unclear, and oats were introduced at about this time. Barley cultivation in one area stopped around AD 700; this was possibly related to isolated and early Viking incursions although soil deterioration may have moved cultivation to another part of Freswick Bay. Pollen and macrofossil analyses have been satisfactorily linked and show how they complement each other. Overall the social and economic changes occurring in the north of Scotland during the first millennium AD are obviously nowhere near as clear as those related to the Romans in northern England. Perhaps Earl's Bu will provide at least some of the missing pieces of the jigsaw.

The hypothesis

Analysis of both carbonized and waterlogged material from Carlisle, Cumbria (Huntley, 1989, 1992a) indicates that the Roman military economy was based upon a mixture of spelt wheat (*Triticum spelta* [nomenclature follows Tutin *et al.* 1964-80]) and barley (*Hordeum* sp.) but with large amounts of fig (*Ficus carica*), grape (*Vitis vinifera*), olive (*Olea europaea*), almond (*Prunus dulcis*), opium poppy (*Papaver somniferum*), coriander (*Coriandrum sativum*) and other taxa characteristic of hot, dry and sunny climates - climates alien to Carlisle. This is, perhaps, not surprising given that the men of the Carlisle garrison were, probably, from the Mediterranean. These Romans did not seem to enjoy simple local fare which, from evidence from a native site within a few kilometres of Carlisle, (Huntley, unpublished) seems to have been barley and little else in the way of culinary herbs and spices although, of course, vegetative material is almost certainly lacking from the archaeological record. The Romans brought their dietary culture with them to the north and, more importantly with respect to this paper, retained it for the 400 years or so that they remained. They did not integrate with the locals in this dietary respect. Can this situation be extrapolated to, for example, the first millennium in northern Scotland with the Viking and Norse incursions from approximately 9th - 11th centuries AD? What effect did they have upon the native economy?

The material

It is surprisingly very rare that truly waterlogged material is found during excavation in many northern Scottish sites; an example of when this did happen is from Warebeth Broch near to Stromness on Orkney (Bell & Dickson, 1989). They showed that, during the 2nd/3rd centuries AD barley was present and had probably been consumed as a broth by the individual whose faeces were preserved. However, on most sites excavated to date in northern Scotland plant remains are preserved only through being burnt. Carbonized material obviously represents a biased picture of what parts of which plants in total were used. Only those likely to come into contact with fire have much of a chance of becoming carbonized although accidents always happen. Someone's fire may get out of hand and burn his house down thus telling the archaeobotanist something about roofing and other house materials. Generally, in Britain, burning means cereals since we have a wet climate and cereals have to be dried prior to safe storage. Thus, for example, peas and beans are likely to be under-represented.

By simple water flotation of large volumes of sediment from, for example, occupation layers or pits uncovered during excavation carbonised material can be separated from the mineral component. The flot is then examined under a microscope and as much as possible of the plant remains identified by comparison with modern reference material of known identification.

Cereal grains can be identified to genus as can much of the chaff, the straw and ear fragments, for example the glumes and rachis fragments; this is a valuable asset for the palaeobotanist. The presence of cereal grains simply implies use but not necessarily local production. However, if large amounts of chaff fragments are present, particularly the culm nodes (unfortunately not identifiable to species) then local production is probable since whole stooks are unlikely to have been transported far.

Before investigating potential changes related to the Viking incursions a baseline is required. Which cereals were in use before the invasions?

The evidence and discussion

Work on Bronze age (500-1500BC) sites in northern Scotland, such as at Lairg (S. Carter & T. Holden, pers. comm) and Suisgill in Helmsdale (van der Veen in Barclay, 1985) have produced large amounts of naked barley (probably 6-row), but with little wheat or oats. By comparison, sites with Pictish material (dating to early to mid centuries AD) produced large amounts of hulled 6-row barley and oats *e.g.* Freswick, Caithness (Huntley, 1992b), likewise Viking and Norse material (dating broadly from the 9th to 15th centuries AD) at Freswick (Huntley *op. cit.)* on the Scottish mainland and Earl's Bu (Huntley, in prep.) and Brough of Birsay (Donaldson & Nye in Morris, 1989) on Orkney produced hulled 6-row barley and oats.

It therefore seems that the plant economy of northern Scotland changed sometime in the few centuries around the birth of Christ from naked to hulled 6-row barley and with the introduction of oats; and that the invasions made no difference to the cereal staple.

If there is no evidence for change in type of cereal preferred is there, nonetheless, an impact upon farming that can be seen? To investigate this aspect pollen analytical work is necessary.

Cereal plants produce pollen which can be identified to the broad types of oats/ wheat, barley (which includes some wild grasses) and rye. Cereal pollen, unlike that of wild grasses, is not dispersed far from the parent plant; thus if cereal pollen is found in the pollen record it means that cereals were probably being grown or processed in the near vicinity, certainly within a few hundred metres. By looking at suitable deposits (peat bogs, lakes) close to archaeological sites and by radiocarbon dating the sediments in these deposits it is possible to investigate cereal cultivation.

At Freswick archaeological investigations showed that settlement had occurred from at least the Pictish period with midden deposits accumulating through to the mediaeval period. Carbonized grain from these midden layers consisted of the hulled 6-row barley and oats with a little bread wheat being found in Norse material (AD 1000 or later). The two main species were in more or less equal amounts. Celtic bean and flax were also found. With respect to the barley and oats no significant differences were shown through time.

A suitable peat deposit was present nearby at the Hill of Harley and was analysed for its pollen (Huntley, in Morris forthcoming). Radiocarbon dating showed that the main peak of barley-type pollen occurred during the first few centuries AD but that little cereal pollen was present during the Viking or Norse periods or, indeed, since then. A low peak did occur in the earlier parts of the 20th century although it was of short duration only.

As stated earlier, barley-type pollen includes grains of wild grasses - particularly coastal species such as *Ammophila arenaria* and *Leymus arenarius*, both of which grow on Freswick Links today. However, the appearance of barley-type grains just after that of pollen of *Plantago lanceolata* and *Plantago major/media*, both taxa generally considered as indicating human activity, along with the coincidence of the peak values of all three taxa suggests that it represents predominantly cultivated barley.

Multivariate analyses of the pollen data did put Chenopodiaceae pollen (which includes both arable and maritime taxa) close to *Hordeum*-type although *Artemisia*, Compositae and Cruciferae, often considered to be arable indicators, were remote. Thus coastal wild grasses may also be represented to a small extent. Interestingly, no *Avena/Triticum* pollen was recorded in spite of several thousand grains of oats being recovered from excavations.

It is therefore suggested that, although barley was initially grown in the area, it was later imported and that oats were imported throughout the first millennium AD. There is, therefore, apparently a change. However, it must be remembered that cereal pollen travels very short distances only and it may be that cereal fields were still present in Freswick Bay but too remote to be represented at Hill of Harley. Other suitable peat deposits were not available to test this hypothesis.

If this is a real change at about AD 700 what could be its cause? Whilst it seems clear that the Viking raids of the 9th century may well have discouraged farming in a coastal area there remain 100-150 years unaccounted for. Evidence for these raids is largely literary and it could be that a single, but devastating, raid at an early stage, about 700 perhaps did occur.

By the 12th century the Earldom of Orkney had become a supreme and organized power although not without struggle. Literary and other environmental evidence suggests that the occupants of Freswick were intensively fishing the seas and, as the area was part of the Earldom, it is tempting to suggest Orkney as a source of the cereals which were perhaps traded for the Freswick fish. Freswick may, therefore, have been a trading outpost, the occupants of which did little in the way of growing their own cereals. The inhabitants may not have even lived at Freswick all the year round.

Excavations at the predominantly Norse site at Earl's Bu, Orphir on Orkney Mainland have produced massive midden deposits currently being analysed. Some of these have filled in the channel of an earlier horizontal mill and therefore there is clear evidence for cereal usage on a grand scale there. There are suggestions of layers in the middens with moderate amounts of barley rachis fragments - suggesting that processing waste was deposited. The site also has late first millennium BC material as well as first millennium AD and if stratigraphy and contexts are secure then a dating programme may elucidate more about changes from naked to hulled barley as well as when oats became important. An adjacent pollen site remains to be analysed.

Acknowledgements

I gratefully acknowledge English Heritage for funding the work on military sites from Carlisle and Carlisle Archaeological Unit for that on the native site. I also particularly thank Dr Colleen Batey, of Glasgow Museums and Art Galleries, and Professor Christopher Morris, of the Department of Archaeology, University of Glasgow, for inviting me to analyse the material from their excavations at Freswick and Orphir respectively and for obtaining funding for this work.

References

Barclay, G.J. (1985). Excavations at Upper Suisgill, Sutherland. *Proceedings of the Society of Antiquaries of Scotland,* **115,** 159-198.

Bell, B. and Dickson, C. (1989). Excavations at Warebeth (Stromness Cemetery) Broch, Orkney. *Proceedings of the Society of Antiquaries of Scotland* **119,** 101-131.

Huntley, J.P. (1989). Plant remains from Annetwell Street, Carlisle, Cumbria: the bulk samples. 103 pp. *Ancient Monuments Laboratory Report* **1/89.**

Huntley, J.P. (1992a). Plant remains from excavations at The Lanes, Carlisle, Cumbria: part 1 - CAL, OGL, OBL and LEL. 201 pp. *Ancient Monuments Laboratory Report* **51/92** (2 parts).

Huntley, J.P. (1992b). Carbonised plant remains, pp. 76-84. In *Excavations at Freswick Links, Caithness, 1980-1982.* Environmental column samples from the cliff-side, pp. 43-102. In *Norse and Later Settlement and Subsistence in the North Atlantic.* (ed. C.D. Morris & D.J. Rackham) (Occasional Paper Series No. 1.) Department of Archaeology, University of Glasgow.

Huntley, J.P. (unpublished). Carbonised plant remains from excavations at the Infirmary site, Carlisle, Cumbria. *Archive Report to Excavator.*

Morris, C.D. (1989). *The Birsay Bay Project. Volume 1. Brough Road Excavations 1976-1982.* University of Durham, Department of Archaeology Monograph series no. 1, 308 pp. Durham: Department of Archaeology, University of Durham.

Tutin, T.G., Heywood, V.H., Burges, N.A., Valentine, D.H., Walters, S.M. & Webb, D.A. (1964-80). *Flora Europaea.* volumes 1-5. Cambridge: Cambridge University Press.

Plants and Vikings: Everyday Life in Viking Age Denmark

D. ROBINSON

Natural Sciences Research Unit, The National Museum, Ny Vestergade 11, 1471 Copenhagen K, Denmark

Summary

The available archaeobotanical evidence from Viking Age sites lying within the boundaries of present day Denmark is summarized and commented upon. The exploitation of both cultivated and wild plant resources is dealt with under the headings - the agrarian economy (crops and weeds), imported plants, collected food plants, brewing, medicinal plants, religion and ritual, wetlands and heaths.

Introduction

The three centuries of the Viking Age make up one of the most colourful, exciting and romantic periods in north-west European history. The very word 'Viking' conjures up a wealth of dramatic images - sleek warships, marauding raiding parties and the richness of Viking art expressed in metal, wood and stone. Everyday life on the other hand has received rather scant attention from archaeologists, historians, storytellers and illustrators. This paper is an attempt to redress the balance, with a look, through the eyes of an archaeobotanist, at Viking daily life in the Viking homelands lying within the borders of present day Denmark. What can the plant remains preserved in postholes, pithouses, middens and wells tell us about the agrarian economy and exploitation of plant resources in the 8th, 9th and 10th centuries? In the following account the available evidence from archaeobotanical investigations, both old and new, will be summarized and briefly discussed.

The archaeobotanical evidence

Plant remains can be preserved in archaeological contexts in a variety of ways, most commonly by charring or waterlogging (Greig, 1989). Less commonly, we find plant remains preserved by desiccation, mineralization, the toxic influence of metals and as impressions in pottery and daub. Results of archaeobotanical analyses are available from a total of 15 Danish Viking Age sites. Fig. 1 shows their location and gives a short description of each of them. The sites are categorized according to the value of the information about crop plants which they provide. It is not possible within the limits of this paper to present a compilation of all the results; reference is therefore made to the original sources (Fig. 1). The 15 sites are unequally distributed across the country, with the majority lying in the northern halves of Jutland and Zealand. In addition we have the important early urban site of Ribe in south-western Jutland and the Ladby ship burial on Funen. There are no analyses from sites in central and southern Jutland, central and southern Zealand and most of Funen, the southern islands or the Baltic island of Bornholm. The picture we can produce is therefore, of necessity, an incomplete one.

From the sites of Gårestrup, Ejstrup, Øster Aalum, Århus Søndervold, Borgergade Horsens, Gammel Lejre and Kregme we have analyses of the fills or floor layers of pithouses. These are small sunken huts characteristic of the late Iron

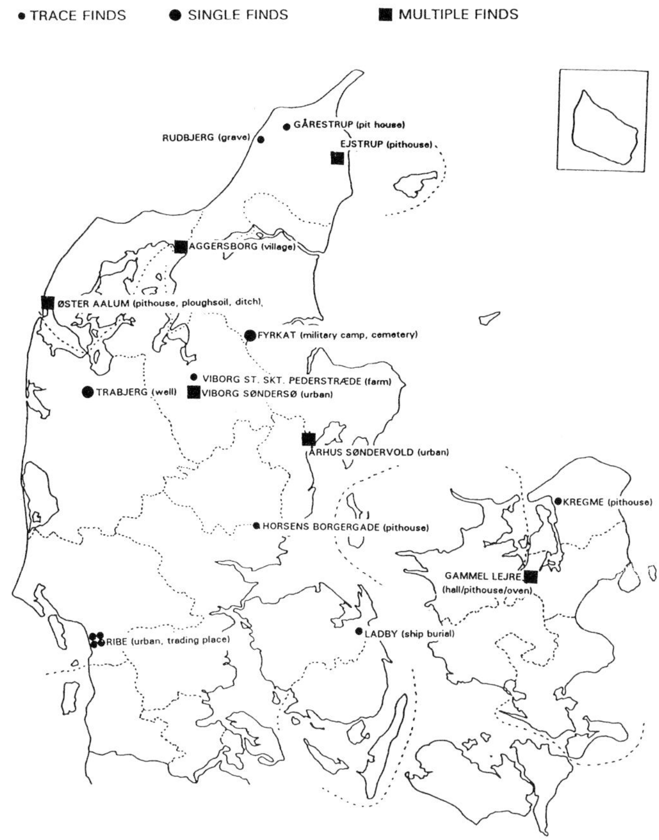

Fig. 1. Archaeobotanical analyses, from Danish Viking Age sites and distribution of finds. Trace finds (•): < 50 crop plant remains; single finds (●): one context with > 50 crop plant remains; multiple finds (■): several contexts with > 50 crop plant remains.

Gårestrup (Robinson, unpublished analyses) charred remains; in fill of pithouse dated archaeologically to late Iron Age/early Viking Age. **Rubjerg** (Robinson, unpublished analyses) charred remains from fill of cremation grave dated archaeologically to early Viking Age. **Ejstrup** (Robinson & Michaelsen, 1989) charred remains from posthole and fill of pithouse radiocarbon dated to AD 885-895 (K-5171). **Aggersborg** (Jessen, 1954) charred remains from rubbish pits belonging to a village predating a military camp from the late 10th century. **Øster Aalum** (Rowley-Conwy, 1988; Robinson & Moltsen, unpublished analyses) charred remains from fill of presumed pithouse radiocarbon dated to AD 780-855 (K-4642), associated plough soil and ditches. **Fyrkat** (Helbæk, 1977) charred contents of a granary in a military camp dating from the late 10th century. Uncarbonized seeds of henbane from a woman's grave in the camp cemetery. **Trabjerg** (Aaby *et al.*, 1992) Predominantly uncharred waterlogged remains from fill of a well, the construction of which is dendro-dated to c. AD 873. **Viborg, St. Skt. Pederstræde** (Jensen, H.A., 1986) Predominantly uncarbonized waterlogged remains from a byre layer dated archaeologically to the late Viking Age. **Viborg, Søndersø** (Robinson *et al.*, 1992) Predominantly uncarbonized waterlogged remains from several contexts including middens, well fills, latrines and floor levels. Dendro-dated to the 11th century. **Århus, Søndervold** (Fredskild, 1971) Predominantly uncarbonized waterlogged remains from peat-like organic deposits and charred remains from fills of pithouses. An early urban site dated archaeologically to the late Viking Age. **Horsens, Borgergade** (Robinson and Moltsen, in press) charred remains from fill of pithouse dated archaeologically to late Viking Age. **Ribe, Kunstmuseets Kælder, Dommerhaven, Tvedgade** (Jensen, H.A., 1986; 1991) & **Posthuset** (Robinson & Boldsen, 1993) four sites within the confines of an early urban trading site in use in the 8th, 9th and 10th centuries (dendro-dates). Low concentrations of charred remains from workshop layers and predominantly uncharred waterlogged remains from boundary ditches. **Ladby ship** (Jessen, 1954) iron encrusted oat grains in a ship burial from the 10th century. **Gammel Lejre** (Robinson, 1993) charred remains from fills of postholes belonging to a 500 m^2 royal hall and the fill of a pithouse, both in use in the 8th and 9th centuries; charred grains from a presumed grain-drying oven radiocarbon dated to AD 660-780 (K-5868). **Kregme** (Robinson & Moltsen, 1992) charred and uncharred remains from fill of pithouse dated archaeologically to the Viking Age.

Age and Viking Period, and are usually interpreted as living quarters or workshops used for weaving, metalwork and the like. The plant remains they contain are almost exclusively charred. At Aggersborg, rubbish pits belonging to a village, demolished to build a 10th century Viking military camp, also contained charred remains, whilst a burnt granary from another military camp at Fyrkat (see Olsen & Schmidt, 1977) produced Denmark's largest Viking Age find of charred grain, a total of 30 l, corresponding to 60 - 70,000 grains, almost all of them rye (*Secale cereale* L.). Charred remains were also in the majority in the fill of a Viking cremation grave at Rubjerg, in the plough soil at Øster Aalum and in the postholes of the impressive royal hall and nearby grain-drying oven, at Gammel Lejre, near Roskilde. In contrast, it was uncharred waterlogged remains which abounded in the well fill from Trabjerg, the midden layers at St. Skt. Pederstræde, Viborg and the various waterlogged urban deposits at Viborg Søndersø and Århus Søndervold. A woman' grave at Fyrkat produced hundreds of uncharred seeds of henbane (*Hyoscyamus),* whilst the plant remains from the Ladby ship burial were restricted to some oat *(Avena)* grains preserved by iron corrosion from the ship's rivets. Finally there is Ribe, reputedly Denmark's oldest town, which began as a trading site early in the 8th century. After levelling the site, using thousands of tons of sand, the area was divided up into narrow lots running at right angles to the river and separated by boundary ditches (see Jensen, S., 1991). Over the following 300 years, these lots housed craftsmen's workshops and traders' stalls, and thick layers of deposits accumulated. The workshop layers were found to contain mostly charred remains, whereas uncharred waterlogged remains abound in the boundary ditches.

The agrarian economy

Crops: Table 1 shows the relative abundance of crop plant remains at sites producing more than 50 remains of crop plants (cf. Fig. 1). Rye, barley (*Hordeum*) (the hulled variety) and, to some extent, oats were clearly the dominant cultivated crops. In addition there are occasional finds of bread wheat (*Triticum aestivum* L.),

Table 1. Summary of single and multiple finds of crop plants from the Danish Viking Age (see also Fig. 1).

SINGLE AND MULTIPLE FINDS (M + S)	BREAD WHEAT	BARLEY UNDIFF.	HULLED BARLEY	NAKED BARLEY	RYE	OATS	BROME	MILLET	FLAX	PEAS
VIBORG SØNDERSØ (M)		***			***	***			*	*
ÅRHUS SØNDERVOLD (M)		***			*****		*		*	*
FYRKAT (S)			*		*****					
AGGERSBORG (M)	*		***	*?	***					
EJSTRUP (M)			***		***	*	*		*	
TRABJERG (S)		*			*****			*	*	
ØSTER AALUM (M)			***		*****	***	*			
GAMMEL LEJRE (M)	*		***	*	***	***	*			

flax (*Linum usitatissimum* L.), millet (*Panicum miliaceum* L.), peas (*Pisum*), naked barley (*Hordeum vulgare* L. var. *nudum*) and brome (probably *Bromus secalinus* L. or *B. arvensis* L.). Species of *Bromus* (in particular *B. arvensis*) have been cultivated in the past but in Viking times there is no evidence to confirm or refute this.

Weed seeds: Table 2 gives information about arable weeds present in the Viking Age. Group one contains species already common in the Iron Age (Helbæk, 1954; Henriksen, 1992). Some of these (*) were collected for food (Helbæk, 1951; Robinson, 1992a) but as yet we have no evidence to confirm that this also was the case in Viking times. Group two contains species which appear at the end of the Iron Age, before increasing, in some cases dramatically, in Viking times. Group three comprises species which appear in the Viking period and become common in the Middle Ages. The majority of species listed in groups two and three are typically weeds of autumn-sown crops. Their appearance and increased abundance reflects the widespread adoption of autumn-sown rye in the course of the Germanic Iron Age and Viking period (Robinson, 1993). Having both autumn- and spring-sown crops was a great advantage, not least because it enabled work in the fields to be staggered through the seasons. Evidence from early mediaeval times suggests that several other changes were associated with this shift in agrarian practice. These include crop rotation with a very short or no fallow period, the creation of plaggen soils, improved ploughs which could turn the soil rather than just disturbing it, and manuring regimes linked to ploughing in long narrow strips and the creation of cultivation ridges (Vejbæk, in press).

Table 2. Weed seeds from Danish Viking Age sites **(* denotes species which were collected for food in the Iron Age.)**

1. WEEDS PRESENT OR COMMON THROUGHOUT THE IRON AGE AND VIKING PERIOD	2. WEEDS PRESENT IN THE IRON AGE WHICH INCREASE IN THE VIKING AGE	3. WEEDS WHICH APPEAR IN THE VIKING AGE AND BECOME COMMON IN THE MIDDLE AGES
Arrenatherum elatius ssp. *bulbosus* *	*Agrostemma githago*	*Anthemis arvensis*
Capsella bursa-pastoris	*Aphanes arvensis*	*Anthemis cotula*
*Chenopodium album**	*Odontites verna*	*Artemisia vulgaris*
*Fallopia convolvulus**	*Sonchus asper*	*Centaurea cyanus*
Fumaria officinalis	*Thlaspi arvense*	*Veronica hederifolia*
*Polygonum lapathifolium**	*Tripleurospermum inodorum*	
*Polygonum persicaria**		
*Polygonum aviculare**		
Raphanus raphanistrum		
Rumex acetosella		
Scleranthus annuus		
*Spergula arvensis**		
Stellaria media		

Imported plants

The only archaeobotanical evidence for imported plants in the Danish Viking Age to date comes from the late 10th century military camp at Fyrkat (Helbæk, 1977). In addition to the earliest Danish records of the imported spice plant coriander (*Coriandrum sativum* L.) and the potherb and medicinal plant bishop's weed (*Aegopodium podagraria* L.) (Jensen, H.A., 1985), the remains of a burnt-down granary contained an enormous find of charred rye grains. Ever since Helbæk's publication of the Fyrkat rye, there has been a great deal of discussion about the origin of the grain. Was it homegrown or imported (Robinson, 1993)? Helbæk was certain that the latter was the case and he gave as his reasons the size of the rye grains, the purity of the find and the presence of exotic contaminants such as coriander and bishop's weed.

Helbæk's conclusions concerning the size and purity of the grain have been questioned by among others Rowley-Conwy (1988). As Rowley-Conwy rightly points out, there have been pure finds of barley and wheat species as early as the Bronze Age (Rowley-Conwy, 1979; 1984) and the Neolithic (Robinson & Kempfner, 1988). With regard to the size of the grains, Rowley-Conwy draws on ethnoarchaeobotanical work on crop processing carried out in the Middle East (Hillman, 1981; 1984). He explains the difference between the Fyrkat rye and other Danish rye finds as being due to crop processing, *i.e.* the Fyrkat rye represents prime grain whereas the other finds comprise a by-product, tail grain, removed by sieving or/or winnowing. The fact remains however that several finds of obviously pure processed rye have been analysed in the intervening period, none of which approach the Fyrkat rye grains in size (Robinson, 1993). It is therefore size, along with Helbæk's third line of argument, the presence of exotic species, which remain the strongest arguments for the Fyrkat rye having been imported. Where the rye came from is another matter; Helbæk favoured eastern Europe.

Collected food plants

Wild food plants have been important dietary supplements at all times in the past, and the Viking Age is no exception. Remains of wild food plants; Hazel (*Corylus avellana* L.), wild strawberry (*Fragaria vesca* L.), crab apple (*Malus sylvestris* Miller), sloe (*Prunus spinosa* L.), raspberry (*Rubus idaeus* L.), blackberry (*Rubus fruticosus* L.) and elder (*Sambucus nigra* L.), are recorded from a number of sites. The Viking Age finds of elder from Kregme (Robinson & Moltsen, 1992), Viborg Søndersø (Robinson, Kristensen & Boldsen 1992) and Ribe (Robinson & Boldsen, 1993) are interesting as they are the earliest Danish records for seeds of this genus (Jensen, 1985). The seeds themselves are very tough and resistant to decay and it is remarkable that they have not been found at earlier sites, particularly in view of the plant's usefulness, not least as a mediccinal plant (Brøndegaard, 1978-80).

Brewing

Both hops (*Humulus lupulus* L.) and bog myrtle (*Myrica gale* L.) are well known as flavouring and preservative agents in beer and remains of both of them have been found in 8th century deposits in Ribe. Bog myrtle is also known from Aggersborg, Viborg Søndersø and St. Skt. Pederstræde. As yet no malted barley grains have been identified, but it seems certain that Danish Vikings know of both hop and bog myrtle beer.

Medicinal plants

When herbals are consulted it appears that almost all plants have been used for medicinal purposes at some time. However, some plants have been used more than others and remains of these have been found at Viking sites:-

A fragment of a seed of greater celandine (*Chelidonium majus* L.) was found at Viborg Søndersø. This plant was used to treat a very wide range of ailments from ringworm to toothache and the plague and was cultivated for the purpose in monastery gardens (Brøndegaard, 1978-80). Elder has also been widely used, in particular for colds and chills, whereas opium poppy (*Papaver somniferum* L.) and henbane were valued for their pain-killing and, in the case of opium poppy, sleep-inducing properties. Flat clubmoss (*Diphasiastrum complanatum* (L.) Holub) and alpine clubmoss (*D. alpinum* (L.) Holub), both found in Ribe, are known as dyeplants, more specifically as mordants (Hall *et al.*, 1984: Robinson, 1986), but they may also have been used medicinally, in particular in treating intestinal parasites as in the case of fir clubmoss (*Huperzia selago* (L.) Bernh. ex Shrank & C. Martius) (Launert, 1981). Similarly, stingining nettle (*Urtica dioica* (L.)), yellow flag (*Iris pseudacorus* L.) and bog bean (*Menyanthes trifoliata* L.) are plants considered to be particularly powerful in treating a number of ills (Brøndegaard, 1978-1980).

The greatest difficulty in discussing the possible medicinal use of plants in the past is providing unequivocal proof. Readily-identifiable residues are very rarely found as the plants are usually ground, boiled or infused before use, and the quantities involved are normally relatively small. However in view of the many ailments from which the contemporary population undoubtedly suffered, and the various written accounts which exist from mediaeval times, it is not unreasonable to presume that plants were also used medicinally in Viking times.

Religion and ritual

Ritual and religious use of plants can also be very difficult to demonstrate, but the Danish Viking Age provides us with some clear examples.

A simple Viking cremation grave at Rubjerg was found to contain grave goods in the form of the charred remains of barley grains and tubers of a subspecies of false oat grass (*Arrhenatherum elatius* L. ssp. bulbosus (Willd.) Spenner) (Fig. 2) (Robinson, unpublished). According to Engelmark (1984) there are several historical accounts of these tubers being used for food in Scandinavia, and charred examples have been found in a number of Iron Age graves (Hjelmqvist, 1955; 1982; Engelmark, 1984; Robinson, 1992b). Engelmark suggests that the tubers may even have been cultivated in much the way that we cultivate potatoes today.

As mentioned earlier, a woman's grave associated with the military camp at Fyrkat was found to contain a collection of several hundred seeds of henbane which had been placed in the grave, possibly in a small leather bag (Helbæk, 1977). Helbæk speculates that they may have been worn as a protection against evil.

When the Ladby ship burial was excavated in the 1930s it was found to contain a great variety of grave goods including the remains of a number of horses and dogs. Among the many finds were some iron-encrusted oat grains which lay next to corroded rivets from the ship's hull (Jessen, 1954). These had apparently been placed in the ship as provisions (for the horses?) on the journey which lay ahead.

Fig. 2. Charred tubers of False Oat Grass (*Arrhenatherum elatius* ssp. *bulbosus*) from a cremation grave at Rubjerg.

Wetlands and heaths

Wetland and heathland species are recorded from the majority of Danish Viking Age sites. They are particularly abundant at early urban waterlogged sites such as Viborg Søndersø, Århus Søndervold and Ribe, where they dominate the plant spectrum (Robinson, in press).

Heathland was an important resource in western Denmark in the past and its influence is clearly seen for example at Viborg Søndersø (Robinson *et al.*, 1992). A well was lined with woven heather (*Calluna vulgaris* (L.) Hull) branches. Heather was also collected for animal fodder and bedding, for fuel, for making brushes and for use in roofing. Heath turf and the underlying peat was very probably also cut for fuel. Bog myrtle, another common plant of the Jutland heath, was sought after for a variety of uses including beer brewing, brush making and as a deodorizer. The latter is hardly surprising given the fact that on a warm summer day the stench from refuse and faeces in Viking Age Viborg would have been overpowering.

Wetland areas were similarly important. The lake-shore vegetation at the time of the earliest Viborg was largely composed of tall-herb species including meadow sweet (*Filipendula ulmaria* (L.) Maxim.), horsetail (*Equisetum*), gypsy-wort (*Lycopus europaeus* L.), bellflower (*Campanula*), yellow flag (*Iris*), club-rush (*Scirpus*), rushes (*Juncus*) and sedges (*Carex*). In wetter areas, and at the waters edge, bogbean and broad-leaved pondweed (*Potamogeton natans* L.) were abundant.

Disturbance of the lake shore by people or cattle allowed plants such as ivy-leaved crowfoot (*Ranunculus sceleratus* L.), which grows on muddy polluted shores, to flourish. In addition to providing the raw material (sedges and rushes) for brushes, ropes, bedding and a range of basket work, the lakeshore vegetation would have also provided grazing for animals and a kind of bog hay for winter fodder (Robinson, 1987).

Conclusion

This brief account reveals that the Vikings made great use of both cultivated and wild plant resources. The agrarian economy was apparently effective, with the use of manure, crop rotation and autumn- and spring-sown crops, but not very diverse, being heavily based on rye, hulled barley and, to a lesser, degree oats. A similar picture is reported from southern Sweden (Engelmark, 1992). It is also clear that, in contrast to the many consumer goods, such as glass, pottery, quernstones and whetstones, which were imported in large numbers even before the Viking Period, plant resources were almost without exception cultivated and collected locally. Possible exceptions to this are however to be found at the Fyrkat military camp.

Plants were also used in medicine, ritual and religion and we can be reasonably confident that beer was brewed (and drunk). Finally, heathlands and wetlands were important as sources of many of the raw materials used in everyday life.

References

Aaby, B., Robinson, D.E. & Boldsen, I. (1992). Pollen- og makrofossilanalyser fra vikingetidsbrønd ved Trabjerg, Ringkøbing Amt. *National Museum, Copenhagen NNU Rapport* **26** (1992), 1-12.

Brøndegaard, V.J. (1978-80). *Folk og Flora,* 1466 pp. Copenhagen: Rosenkilde and Bagger.

Engelmark, R. (1984). Two useful plants from Iron Age graves in central Sweden. *Archaeology and Environment* **2**, 87-92.

Engelmark, R. (1992). A review of the farming economy in South Scania based on botanical evidence. In *The archaeology of the cultural landscape: Acta Archaeologica Lundensia* 4(19) (ed. L. Larsson. J. Callmar & B. Stjernquist), pp. 369-376. Stockholm: Almqvist & Wiksell.

Fredskild, B. (1971). Makroskopic planterester fra det ældste Århus. In *Århus Søndervold: en byarkæologiske undersøgelse. Jysk Arkæologisk Selskabs Skrifter* IX (ed. H.H. Andersen, P.J. Crabb & H.J. Madsen), pp. 307-318. Copenhagen: Nordisk Forlag.

Greig, J.R.A. (1989). *Archaeobotany. Handbooks for Archaeologists No4,* 93 pp. Strasbourg: European Science Foundation.

Hall, A.R., Tomlinson, P.R., Hall, R.A., Taylor, G.W. & Walton, P. (1984). Dyeplants from Viking York. *Antiquity* **58,** 58-60.

Helbæk, H. (1951). Ukrudtsfrø som næringsmiddel i førromersk jernalder. *KUML* ***1951***, 65-74.

Helbæk, H. (1954). Prehistoric food plants and weeds in Denmark. A survey of archaeobotanical research 1923-1954. *Danmarks geologiske Undersøgelse* **(II)80**, 250-261.

Helbæk, H. (1977). The Fyrkat grain: a geographical and chronological study of rye. In *Fyrkat. En jysk vikingeborg I. Borgen og bebyggelsen. Nordiske Fortidsminder* **(B)3** (ed. O. Olsen & H.W. Schmidt), pp. 1-41. Copenhagen: Det Kgl. nordiske Oldskriftselskab.

Henriksen, P.S. (1992). Jernalderens landbrug beskrevet ud fra arkæologiske frøfund. *National Museum, Copenhagen NNU Rapport* **20 (1992)**, 1-16.

Hillman, G.C. (1981). Reconstructing crop husbandry practices from charred remains of crops. In *Farming practice in British Prehistory* (ed. R. Mercer), pp. 123-162. Edinburgh: University Press.

Hillman, G.C. (1984). Interpretation of archaeological plant material. The application of ethnographic models from Turkey. In *Plants and Ancient Man* (ed.W. van Zeist & M.A. Casparie), pp. 1-42. Rotterdam: Balkema.

Hjelmqvist, H. (1955). Die älteste Gesichichte der Kulturpflanzen in Schweden. *Opera Botanica* **1**(3), 1-186.

Hjelmqvist, H. (1982). Arkeologisk botanik - något om metoder och mål. *Svensk botanisk tidskrift* **76**, 229-240.

Jensen, H.A. (1985). *Catalogue of late- and post-glacial macrofossils of Spermatophyta from Denmark, Schleswig, Scania, Halland and Blekinge dated 13000 B.P. to 1536 A.D. Danmarks Geologiske Undersøgelse Serie A, no. 6.*, 95 pp. Copenhagen: C.A. Reitzals Forlag.

Jensen, H.A. (1986). Seeds and other diaspores in soil samples from Danish Town and Monastery Excavations, dated 700-1536 A.D. *Biologiske Skrifter* **26,** 1-107. Copenhagen: Det Kongelige Danske Videnskabernes Selskab.

Jensen, H.A. (1991). Seeds and other macrofossils in the 8th century deposits. In *Ribe excavations 1970-76 volume 3* (ed. Bencard, M., Jørgensen, L.B. & Madsen, H.B.), pp. 17-36. Esbjerg: Sydjysk Universitetsforlag.

Jensen, S. (1991). *Ribes Vikinger.* 72 pp. Ribe: Den antikvariske Samling.

Jessen, K. (1954). Plantefund fra vikingetiden i Danmark. *Botanisk Tidsskrift* **50,** 125-139.

Launert, E. (1981). *Edible and Medicinal Plants of Britain and Northern Europe.* London.

Olsen, O. & Schmidt, H.W. (1977). *Fyrkat. En jysk vikingeborg I. Borgen og bebyggelsen. Nordiske Fortidsminder* **(B) 3,** 241 pp. Copenhagen: Det Kgl. nordiske Oldskriftselskab.

Robinson, D.E. (1986). Clubmosses from medieval Perth. *Antiquity* **60,** 49-50.

Robinson, D.E. (1987). The botanical remains. In *Excavations in the medieval burgh of Perth 1979-1981* (ed. P. Holdsworth), pp. 199-209. Edinburgh: Society of Antiquaries of Scotland.

Robinson, D.E. (1992a). Et fund af forkullede ukrudtsfrø fra Borremose. *National Museum, Copenhagen NNU Rapport* **29 (1992),** 1-5.

Robinson, D.E. (1992b). Vinding - makrofossilanalyse af brandgrav fra ældre romersk jernalder. *National Museum, Copenhagen NNU Rapport* **24 (1992),** 1-5.

Robinson, D.E. (1993). Plant remains from the late Iron Age/early Viking Age settlement at Gammel Lejre. *Journal of Danish Archaeology* **10,** 191-198.

Robinson, D.E. (in press). Botanical investigations at Danish urban sites - a review of the past and a look to the future. *Hikuin.*

Robinson, D.E. & Boldsen, I. (1993). Botaniske analyser af prøver fra udgravningen ved Ribe Posthus. *National Museum, Copenhagen NNU Rapport* **32 (1993),** 1-16.

Robinson, D.E. & Kempfner, D. (1988). Carbonised grain from Mortens Sande 2. *Journal of Danish Archaeology* **6 (1987),** 125-129.

Robinson, D.E., Kristensen, H.K. & Boldsen, I. (1992). Botanical analyses from Viborg Søndersø: a waterlogged urban site from the Viking period. *Acta Archaeologica* **62 (1991),** 59-87.

Robinson, D.E. & Michaelsen, K.K. (1989). Early Viking Age carbonised plant remains from Ejstrup, Vendsyssel, Denmark. *Acta Archaeologica* **59 (1988),** 223-228.

Robinson, D.E. & Moltsen, A. (1992). Planterester fra et vikingetids grubehus ved 'Birkely', Kregme i Nordsjælland. *National Museum, Copenhagen NNU Rapport* **23 (1992),** 1-9.

Robinson, D.E. & Moltsen, A. (in press). Botaniske analyser af prøver fra det tidlige Horsens 1991-92 - Borgergade, Nørregade, Søndergade og Kirketorvet. *National Museum, Copenhagen NNU Rapport.*

Rowley-Conwy, P. (1978). Forkullet korn fra Lindebjerg. *KUML* **1978,** 159-171.

Rowley-Conwy, P. (1984). The Egehøj cereals. Bread wheat (*Triticum aestivum* s.l.) in the Danish Early Broze Age. *Journal of Danish Archaeology 3,* 104-110.

Rowley-Conwy, P. (1988). Rye in Viking Age Denmark: new information from Øster Aalum, North Jutland. *Journal of Danish Archaeology* **7,** 182-190.

Vejbæk, O. (in press) Gødningsrug og sandrug: Vintersæd og agertyper i Danmarks yngre jernalder og tidlige middelalder.

Exploitation of Local, Drifted and Imported Wood by the Vikings on the Faroe Islands

C. MALMROS

Natural Sciences Research Unit, The National Museum, Ny Vestergade 11, 1471 Copenhagen K, Denmark

Summary

Two hundred wood samples have been examined from the Viking settlement at Argisbrekka on Eysturoy dated to AD 770-1015. They fall into three categories:-1. local wood: *Juniperus, Betula pubescens* Ehrh., *Corylus, Salix* and heathland dwarf shrubs; 2. driftwood: *Picea, Larix* and *Pinus* section *strobus;* and 3. imported wood: *Quercus, Pinus* section *sylvestris, Alnus* and *Fraxinus.* The local tree vegetation was over-exploited and *Betula* became extinct in the course of the Viking Age. Driftwood, mainly of Siberian origin, seems to be the most important tree resource and was used for houses and utensils. Imported wood probably mostly comprised ships and finished products made in Scandinavia. The detection of tree *Betula* growth between 2,460 BC and AD 770 is a new contribution to the vegetational history of the Faroe Islands.

Introduction

The present vegetation on the Faroe Islands in the North Atlantic Ocean between Norway and Iceland is strongly influenced by the presence of thousands of sheep which graze grassland and dwarf-shrub heaths dominated by *Calluna* and *Empetrum.* Native trees are not found but in a few places, inaccessible to sheep, low bushes of *Juniperus communis* L. ssp. *nana* (Willd.) Syme, *Salix glauca* L. and *S. phylicifolia* L. may occur. The production from the few plantations is too small to provide the Faroese inhabitants with timber and quantities of wood are imported for construction, house and ship building. Up to the end of the last war, utilization of driftwood was a common practice, the importance of which is demonstrated by old rules regarding the ownership of stranded wood. Analyses of wood from a Viking Age settlement on Eysturoy show that the Norsemen who settled on the islands around AD 800, found native trees which were exploited together with driftwood and imported wood (Malmros, 1990).

The excavations at Argisbrekka

The Viking settlement of Argisbrekka lies on the northern part of Eysturoy, 1.5 km inland from the coast and 130 m above sea level, at the eastern end of Lake Eidisvatn (Mahler, 1991). A small river, running from the high mountains, has formed an alluvial plain, a delta, which before the construction of a hydro-electric plant in 1988, was covered by grass and was a fine pasture for sheep and cattle.

The excavations showed the course of the Viking Age stream with remains of 17 houses lying on its banks. In the sediments of the former watercourse, refuse from the settlement and material from disused houses was found mixed up with secondarily deposited material from pre-Viking peat layers eroded by the stream.

A section in the old peat layers revealed that 0.6 m below the surface there was a 0.25 m thick layer of wood peat with roots, stems, branches and twigs of *Betula cf. pubescens* and *Juniperus* resting on peat with *Calluna*. The deposit is probably identical with a locality of buried trees mentioned by the Scots geologist James Geikie (1880):'...at Eide in Osteroe, the merchant there told me he had seen them taken from the peat near Eide Vatn as thick as his leg. I could not determine the species of wood with any certainty, but pieces I saw were probably juniper and birch.'

Objects of wood constitute the overwhelming part of the archaeological material. In the houses, only charcoal, some hazel nuts and a few fragile pieces of decayed wood and bark were found. In the waterlogged refuse layers in the river bed the state of preservation, especially of the wood of conifers and dwarf shrubs, was much better due to the constantly cold, wet and anaerobic conditions; wood of larger broad-leaved trees was generally more decayed.

Radiocarbon dates

Sixteen samples of wood and nutshells from the excavation have been dated in the radiocarbon dating laboratories in Copenhagen and Uppsala (Fig. 1). In the text only the calibrated mean values are cited. The dates illustrate some of the problems commonly found when dating archaeological sites and Holocene deposits in a mountain area like the Faroe Islands. The radiocarbon dates of wood sometimes are surprising, but when the origin of wood and the stratification of the site are understood every date can be satisfactorily explained.

Organic sediments were omitted for radiocarbon dating because the evaluation of the results would be much more difficult. Not only gyttja but also peat layers were suspected of containing old re-deposited material due to the often high content of clay, sand and gravel indicating erosion of the mountain sides beyond Argisbrekka.

Samples of local wood, hazel nuts and imported wood found in the houses indicate the settlement period: AD 770-1015, while samples of driftwood of *Picea* and *Larix* could be up to 200-350 years old when used. Wood samples from the fossil river showed dates ranging from 1045 BC to AD 1010, which demonstrates the composite nature of the sediments.

Wood analysis

Only a minor part of the large amount of preserved wood has been examined, comprising 109 pieces of unworked wood, 38 artefacts and 54 pieces of carpentry waste etc. (Fig. 2). Fifteen taxa have been identified, these can be divided in three groups of different origin: local wood, driftwood and imported wood.

Local wood

Eight taxa have probably grown at Argisbrekka: *Juniperus, Betula, Salix* and *Corylus* and heathland dwarf shrubs: *Calluna, Empetrum, Vaccinium myrtillus* L. and *Vaccinium uliginosum* L.

Juniperus was used for ropes, bucket loops, and short pointed sticks. The wood is common on Viking Age sites and has been used until a few generations ago (Larsen, 1991). Fossil *Juniperus* is well known from the peat bogs on the Faroe Islands and this seems to be the low northern form *Juniperus communis* ssp. *nana*(Jóhansen, 1985). According to the radiocarbon dates, the species grew at Argisbrekka from the middle of the 3rd century BC to the beginning of the 11th centuryAD.

Worked wood of *Betula* has not yet been identified, but rolls of bark which could be used as roofing material were found in a house; probably a great deal of wood was used as fuel for domestic purposes and iron-smelting but charcoal analyses are lacking. Four radiocarbon dates show that *Betula pubescens* was present in the area over a period of more than 3,000 years. The oldest sample, dated to 2,460 BC, is a 0.12 m thick stem from the fossil layer of wood peat and the youngest a bark roll from AD 770. More recent *Betula* wood is recorded from other Viking settlements on Eysturoy, at Eide (radiocarbon date AD890 K-5618) and (Toftanes 9-10th century AD (Hansen, 1991).

A small model boat was made of a branch of *Salix* probably *S. glauca* or *S. phylicifolia,* the only native willow trees occurring on the Faroe Islands today.

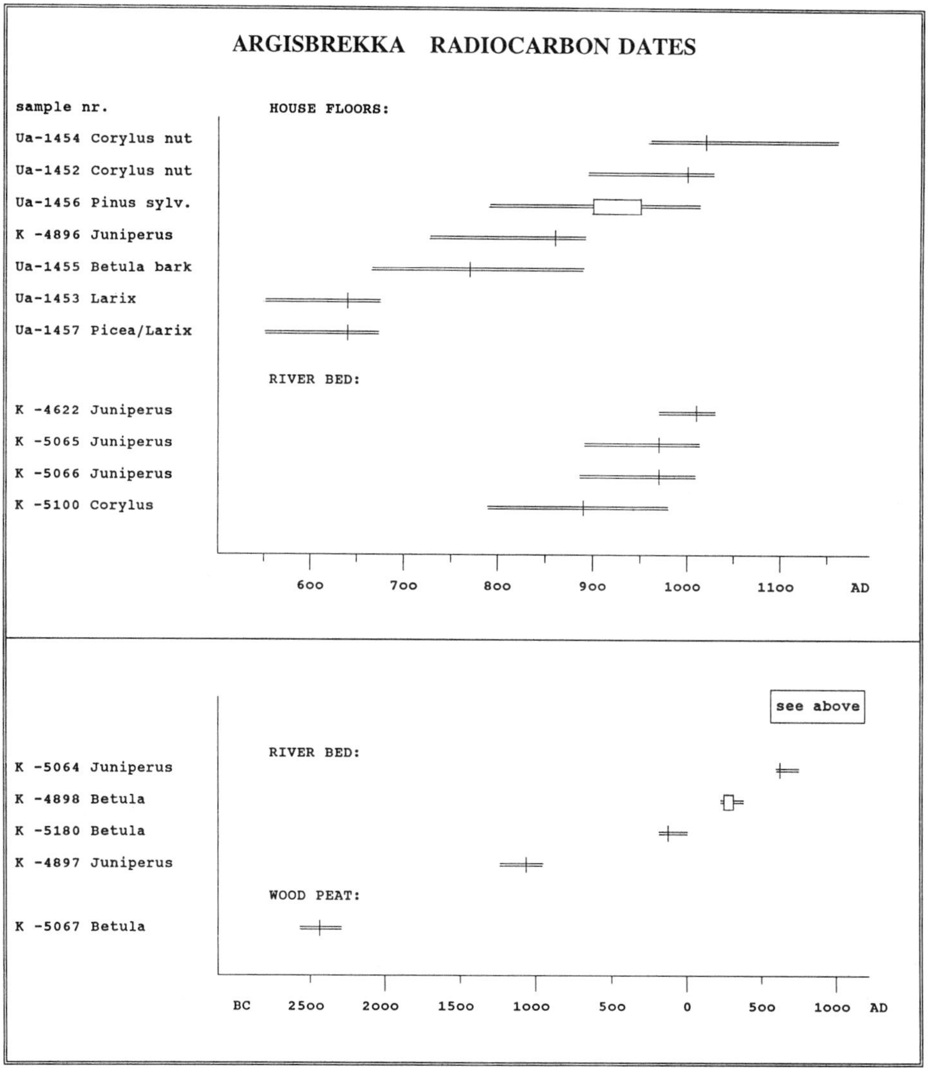

Fig. 1. Radiocarbon dates from the excavation at the Viking Age settlement at Argisbrekka, Faroe Islands. K-nr.: Copenhagen conventional dating. Ua-nr.: Uppsala AMS dating. Calibration ± 1 standard deviation (Pearson & Stuiver, 1986; Stuiver & Pearson, 1986).

Corylus wood is represented by a 0.08 m thick trunk with basal cutmarks from an axe indicating felling. The trunk was found in the refuse layer and has a radiocarbon age of AD 890. In addition, nuts of *Corylus avellana* occur on the house floors and in refuse layers and they are also known from Toftanes and other Viking sites on the Faroe Islands. These nutritious nuts could have been imported from Scandinavia or Britain but the felled tree trunk indicates possible local growth in the Viking Age, when the climate was milder than at present. The immigration of *Corylus* seem to be related to human activity as the large nuts could not naturally spread across the sea, a problem discussed by Boyd & Dickson (1986).

ARGISBREKKA WOOD ANALYSIS

		A	B	C	D
Local wood:					
Juniper	*Juniperus communis*	79	6	..	85
Tree birch	*Betula pubescens*	11	..	2	13
Willow	*Salix sp.*	..	1	..	1
Hazel	*Corylus avellana*	..	..	1	1
Local dwarf shrubs:					
Heather	*Calluna sp.*	10	1	..	11
Crowberry	*Empetrum sp.*	6	..	..	6
Bilberry	*Vaccinium myrtillus*	1	..	..	1
Bog whortleberry	*Vaccinium uliginosum*	2	..	..	2
Driftwood:					
Spruce	*Picea sp.*	..	4	5	9
Larch	*Larix sp.*	..	8	9	17
Spruce or Larch	*Picea/Larix*	..	1	24	25
White pine	*Pinus* section *strobus*	..	..	1	1
Imported wood:					
Scots pine	*Pinus* section *sylvestris*	..	12	8	20
Oak	*Quercus sp.*	..	3	4	7
Alder	*Alnus sp.*	..	1	..	1
Ash	*Fraxinus sp.*	..	1	..	1
Sum:		**109**	**38**	**54**	**201**

Fig. 2. Analysis of wood from the excavation at the Viking Age settlement at Argisbrekka, Faroe Islands. A: number of unworked wood. B: artefacts. C: carpentry waste etc. D: total number.

As *Corylus* is sparse in pollen diagrams from the Faroe Islands (Jóhansen, 1985) and the tree is unknown from historical times, possible cultivation must therefore have been rather restricted in time and distribution.

Calluna is used for cordage. This and the other species of dwarf shrubs *Empetrum, Vaccinium myrtillus* and *Vaccinium uliginosum* are still found in heathland on the Faroe Islands.

Driftwood

This group comprises three taxa: *Picea, Larix* and *Pinus* section *Strobus*. Only one piece of the latter has been recorded. A separation of *Picea* and *Larix* is not always possible on the basis of their anatomy, and an identification to European (Siberian) and North American subspecies of *Picea* and *Larix* has not been attempted because the characters normally used are variable (Phillips, 1948; Schweingruber, 1990). *Pinus strobus* L. has anatomical features in common with *Pinus cembra* L. which grows in central Europe and is certainly not found in Atlantic driftwood.

The presence of holes bored by ship worms prove the origin of *Picea* and *Larix* as driftwood. Similar holes are not recorded in samples of *Pinus* section *sylvestris,* but do not exclude the possibility that driftwood of pine may occur. For other reasons it is here grouped under imported wood.

At Argisbrekka the raw timber has been exclusively worked by cleaving and cutting with axe and knife, while the saw was probably not used. Some surfaces are very smooth and may be finished by a scraping tool or plane. The artefacts include wooden dowels, short pointed sticks, ownership tags, toggles. To this can be added many pieces of waste from house construction and manufacturing of artefacts.

Since the Faroe Islands are located in a region where the East Icelandic Polar Current and the North Atlantic Current meet it is possible to find wood of North European as well as of East American origin. Only one piece of *Pinus* section *Strobus* has been detected from Argisbrekka, and it is therefore presumed that most of the driftwood which accumulated on the northern beaches of the Faroe Islands in the Viking Age came with the East Icelandic Polar Current.

In the Polar Sea and North Atlantic Ocean huge amounts of Siberian driftwood are transported from east to west by the Polar Current and wood of *Picea* and *Larix* and a smaller portion of *Pinus sylvestris* is found in recent deposits on the beaches on Svalbard, Jan Mayen, Iceland, East and West Greenland (Ingvarson, 1903).

The North Atlantic Current which runs from west to east transports North-east American driftwood to Europe. On the Hebrides and Northern Isles of Scotland, this current must be responsible for North American driftwood found on archaeological sites ranging in age from the Neolithic to the 16th century AD. (Dickson, 1992).

Some driftwood circulates in the Polar Sea and the North Atlantic. In two areas, south of the Denmark Strait and in the North-east Atlantic, the Polar and the North Atlantic currents meet and cold and warm water become partially mixed. Some driftwood is exchanged and transported in the opposite direction. This explains the presence of small amounts of North American driftwood found on the coasts of the Polar Sea as well as Siberian driftwood on the coasts of Western Europe.

Samples of driftwood can present serious problems when radiocarbon dating archaeological sites because it is not known how long the wood was at sea and how long it had lain on the beach before being collected by man (Mahler & Malmros, 1991). Two samples from Argisbrekka are both dated to AD 640. One sample of *Larix* (Ua-1453), is from the same house as a 360 years younger *Corylus* nut sample (Ua-1452), and the other sample of *Picea* or *Larix* is around 200 years older than indicated by its stratigraphic position.

More recent driftwood has been dated from Toftanes; a sample of *Picea* or *Larix* has a calibrated mean value of AD 900-955 (K-4441) which corresponds to the period from the 9th to the 10th century of the site (Malmros, 1990).

Imported wood

This group comprises four taxa: *Pinus* section *Sylvestris, Quercus, Alnus* and *Fraxinus. Pinus* was commonly used for small tools and buckets and was worked in the same way as the driftwood.

The number of hardwood items is rather small, a fragment of a boat plank with iron rivets and the bottom of a bucket are identified as *Quercus*, and carpenter's waste shows that oak was probably incorporated into house constructions. A turned plate of *Alnus* and a knife handle of *Fraxinus* most probably arrived at Argisbrekka as finished products. The latter is re-used as a counting stick probably at lambing time.

All of the four taxa mentioned are native to Western Norway, a region with which the Vikings on the Faroe Islands had the closest contact. The wood most probably came from there together with other goods. The wood may have arrived as timber or finished products, for instance in the form of ships and boats, for which the use of *Quercus* and *Pinus sylvestris* is well documented. When no longer seaworthy they represented a valuable resource, and the boat plank mentioned may indicate re-use of wood from a demolished vessel.

A radiocarbon sample of *Pinus* section *Sylvestris* found in a house is dated to AD 900-950 (Ua-1456) and falls within the range of dates for local wood from the settlement period.

Conclusion

The preliminary analysis of the wood material from Argisbrekka illustrates the wood resources exploited by the Vikings on the Faroe Islands in the 9th to 11th century AD. When man arrived he found a local vegetation quite different from that of the present. Trees of *Betula pubescens* and bushes of *Juniperus communis* were common in the area and had been present for more than 3,000 years. The find of *Betula* wood confirms Geikie's observation from 1880 and makes a new contribution to the vegetational history of the Faroe Islands. According to Jóhansen (1985) tree *Betula* was presumed never to have immigrated to the islands, although it is well known from the Shetlands and Iceland.

Local wood resources were over-exploited by the Viking settlers and regrowth was prevented by agriculture and the grazing of farm animals. *Betula* became extinct in the course of the Viking Age while *Juniperus* and *Salix* were forced back into restricted areas. *Corylus* was introduced as a food crop, but cultivation only lasted for a relatively short period.

Driftwood of Siberian origin was the most important wood resource and especially *Picea* and *Larix* were used for houses and artefacts and probably also for ships and boats, as on Iceland and in the Norse settlements in West Greenland (Andersen & Malmros, 1993). Imported Scandinavian wood played a minor role at Argisbrekka and probably mostly arrived on the Faroe Islands as sailing vessels and as smaller finished products.

Wood was of fundamental importance for the inhabitants on the Faroe Islands and the way in which they exploited the different resources, corresponds to the practices met with in the later Norse settlements in the North Atlantic region. Pollen analyses show a marked decline in tree *Betula* pollen frequencies at the time of the *Landnám* in Iceland at about AD 900 and in South Greenland at AD 1,000, and the decline is explained as the result of severe human impact on the woodlands, followed by grazing and erosion (Hallsdóttir, 1987; Sandgren & Fredskild, 1991). A great dependence on driftwood resources is demonstrated by the well-preserved remains of buildings, ships and artefacts found on Norse settlements in Greenland.

References

Andersen, E. & Malmros, C. (1993). Ship's parts found in the Viking settlements in Greenland. Preliminary assessment and wood-diagnoses. In *Viking Voyages to North America* (ed. B.L. Clausen), pp. 118-122. Roskilde: Vikingeskibshallen.

Boyd, W.E. & Dickson, J.H. (1986). Patterns in the geographical distribution of the early Flandrian *Corylus* rise in southwest Scotland. *New Phytologist* **102,** 615-623.

Dickson, J.H. (1992). North American driftwood, especially *Picea* (spruce) from archaeological sites in the Hebrides and Northern Isles of Scotland. *Review of Paleobotany and Palynology* **73,** 49-56.

Geikie, J. (1880). On the geology of the Faroe Islands. *Transactions of the Royal Society of Edinburgh* **30,** 217-269.

Hallsdóttir, M. (1987). Pollen analytical studies of human influence on vegetation in relation to the Landnám tephra layer in southwest Iceland. *Lundqua Thesis* **18.**

Hansen, S. Stummann (1991). Toftanes: A Faroese Viking Age farmstead from the 9-10th Centuries A.D. *Acta Archaeologica* **61-1990,** 44-53.

Ingvarson, F. (1903). Om drifveden i Norra Ishavfet. *Kongelige Svenska Vetenskaps-Akademiens Handlingar* **37**(1).

Jóhansen, J. (1985). Studies in the vegetational history of the Faroe and Shetland Islands. *Annales Societatis Scientiarum Færoensis Supplementum* **XI.**

Larsen, A.-C. (1991). Norsemen's use of juniper in Viking Age Faroe Islands. *Acta Archaeologica* **61-1990,** 54-59.

Mahler, D.L.D. (1991). Argisbrekka: new evidence of shielings in the Faroe Islands. *Acta Archaeologica* **61-1990,** 60-72.

Mahler, D. & Malmros, C. (1991). Comments on the early settlement of Iceland. *Norwegian Archaeological Review* **24**(1), 14-18.

Malmros, C. (1990). Viking Age wood resources at Argisbrekka, Faroe Islands. *Norwegian Archaeological Review* **23**(1-2), 86-92.

Pearson, G.W. & Stuiver, M.C. (1986). High precision calibration of the radiocarbon time scale, 500-2500 BC. *Radiocarbon* **28** (2B), 839-862.

Phillips, E.W.J. (1948). Identification of softwoods by their microscopic structure. *Forest Products Research Bulletin* **22.**

Sandren, P. & Fredskild, B. (1991). Magnetic measurements recording Late Holocene man-induced erosion in S. Greenland. *Boreas* **20,** 315-331.

Schweingruber, F.H. (1990). *Anatomie Europäischer Hölzer,* 800 pp. Stuttgart: Haupt.

Stuiver, M.C. & Pearson, G.W. (1986). High-precision calibration of the radiocarbon time scale, A1950-500 BC. *Radiocarbon* **28** (2B), 805-838

Habent Fata Sua: The Rise and Fall of Garden Plants

K. FAEGRI

Botanical Institute, Allégaten 41, University of Bergen, Norway

Special Guest Lecture

'Habent fata sua'...these words, originally a reflection on the fickleness of human literary preference, apply also to the relation between man (including, of course, woman) and plants, more especially flowers, flowers for enjoyment and for pleasure, the flowers of our garden beds and flower-pots. It is a fascinating theme above all because of the speed with which changes take place. It may take decennia before a new fruit or vegetable is accepted - the history of the potato being the classical one. New species are considered with deep distrust and often not accepted until dire necessity forces them upon us. A new garden flower, on the other hand, is accepted instantly. Present a good one, and people cannot wait to get hold of it - *per fas et nefas* - for themselves. Changes in this sphere of plant relations are very fast. The evaluation of a new strain of vegetables or fruit depends on easily defined, measurable factors: ease of cultivation, quantity of produce, quality of produce, larger, sweeter, more easily harvested. Flowers are also, but to a lesser extent, judged by similar criteria, but there is a major factor: fashion, and the fickleness of fashion needs no commentary.

Even from prehistoric times there are suggestions that man related to plants, flowers, also for sentimental reasons, and after the advent of chronicling - visual or by letters - there is ample evidence of the same. The Egyptian gardens are well documented in murals. The Greeks seem, on the whole, to have been satisfied with the immense richness of their spontaneous flora (their gardens were mostly orchards), but the Romans again, almost to excess, used flowers to adorn themselves and their surroundings. They could and did draw on the wealth of flowers of the whole empire, the Near East and from even further away.

With the decline and fall all this came to an end. The barbaric hordes hardly had any use for flowers, and with the barbarizing of the Roman empire, western horticulture fell upon lean times. During the Dark Ages monasteries formed the only narrow passage through which part of the Classical inheritance - severely cropped by the Church - passed on to a more affluent and broad-minded posterity. At any rate secular enjoyment certainly had no place in the program of ecclesiastical institutions. The monks cultivated their plants for practical uses, food or medicine, and thanks to the ideal monastery plan from St. Gallen we know a great deal about how it should properly be done. The early Church (and puritan movements throughout the history of Christianity) had no use for flowers, considering them with distrust as being part of pagan cults - a suspicion that was not unfounded. The dilemma was solved in the traditional way, accepting the form and filling it with Christian symbolism.

Those Classical garden plants that came down to us certainly came the apothecary way, but if a medicinal plant just happened also to be a thing of beauty it was in itself not catastrophic. Indeed, one may wonder if the medical properties were not in some cases subordinate to the aesthetic. Roses, lilies, irises, violets had medical reputations, but was it much more than an excuse? When the abbot of a monastery in south-east Norway got a silver rose as a present, it is difficult to imagine that it only reflected the medical virtues of that flower.

We have a more direct testimony in the touching Hamar Chronicle, one of the most human documents that have come down to us from the 16th century. The monastery and cathedral of Hamar were sacked in the early 1500s first by the Swedes and later, through the Reformation, by the King. On the occasion of the final disbanding of the ecclesiastical institutions, one of the monks - we like to think that it was old Trugels Precentor, 'almost a hundred years old' - wrote this little Chronicle from which I quote:

> 'And was it very blithesome [old Norse word order] in the summer when you went a-rowing at Hamar town [in the lake], as all herbs and trees gave off such wonderful scent. And the pilgrims who went to Rome and the Holy sepulchre, and through many places and countries, they did their utmost to bring back on the way back to their homeland some of the fine scented herbs that were dearest to them and did they at that time in Hamar especially love a tree that was called Angels' thorn [sweet briar to-day] and also gave off a wonderful fragrance.'

It is touching reminiscence of the old days and I suspect that Old Trugels' joy at his garden was not exclusively ecclesiastical and religious.

Symbolism became a very important aspect of man's contact with flowers, nurtured and maintained, above all, by the Church after it had accepted flowers in a religious context. Some of these symbols are obvious like the (very late) symbolism of the passion flower. Others are of great age and less obvious. Why myrtle just for innocence? The white lily also stands for the same - virginity - as shown by innumerable white lilies as obligate adjuncts to paintings of the Annunciation. Do not we still have a feeling of solemnity and purity seeing one of those gorgeous blossoms which now have changed into a symbol of death?

There was the rose - for love, which I presume was intended to represent *agape*, the spiritual love. However, inevitably Eros - carnal love - got his fingers into the pie with sundry non-ecclesiastic results. I know of an early 19th century great poem - one of the greatest in Norwegian literature (nobody reads it!) - in which the similarity of a moss rose bud with certain parts of the female anatomy is so explicitly expressed that I wonder if even a modern poet would have dared to use plain words - that is, if you know the buds of those long-forgotten roses. And there was rue for repentance - but you certainly know your Ophelia much better than I do, and there is no reason to continue.

The esoteric language of symbols was a characteristic of the Church, and with the waning influence of the Roman Catholic Church at the Reformation the symbolism lost its spiritual basis and is today a thing of the past even if many of the species still linger and we have faint remembrances due more to convention than to real understanding of the underlying motifs.

However, this Don Quixote also had his Sanco Pansa, the 'secret' language of flowers by means of which 19th century couples expressed feelings they could not/ would not/ dared not express directly. Various flowers communicated warm feelings, sometimes rather complicated ones, like bilberry leaves in Norway 'I don't promise very much, but I keep my promises'. This language is not recorded before 1818 (English translation in 1834), which is a very interesting date. In the previous centuries, although romantic love was far from unknown, one has the feeling that

Decamerone was closer to reality than Abelard and Heloise, plus some solid economic background ideas. Perhaps such complicated expressions of romantic love simply were not necessary before the age of the Young Werther and his co-sufferers. On the practical side: mail was a very complicated and expensive commodity that certainly was not wasted wantonly. Both of these features changed around 1800. Romanticism came out - excessively so, - and mail services developed gradually. Whereas the mail service is still there (let's forget the quality) the need for such communications has certainly vanished and young people express their feelings more directly. The flower language was still alive, only just, when I was a child. Now it is completely dead, even more so, than romanticism.

Former periods were more intrigued by abnormalities, oddities, in the floral world than we are. In some respects this interest was rather morbid. In plants, matters were more idyllic. A white-flowered variety of a usually red or blue flower, a fasciation, a cut-leaved strain, black flowers (*Iris susiana*) and, above all, double flowers were highly appreciated. They were the fashion and the more the better. Flowers were filled with extra petals to and beyond the beauty limit. In old nursery lists and seedsmen's catalogues there are no end of double flowers, and there is hardly a single genus of cultivated flowers which is not represented also by 'doubles' - most of which have now been lost. A few survive as garden oddities, but in very few genera this trend towards filled flowers is maintained. Exceptions are flowers used in the perfume industry and having petals as osmophores (scent-emitting organs). For them the more the better.

The role of odours in man's dealing with flowers, wild and in the garden, always intrigues me. Apparently, there are two entrances to this theme. One is the medicinal one. A majority of herbs used medically exhaled a strong fragrance, which may or may not have indicated the presence of therapeutically effective substances. Even if the herbs had no medical effect as such, the scent certainly had a psychological effect. The old monastic *materia medica* was, practically without exception, strongly fragrant. The way *via* the monastery garden was so important for the survival of the garden flora. Of yore, scent was one of the passwords on that journey. One suspects that in many cases the plant remained on the medicine list for its fragrance long after the medical role was forgotten.

The other route is more earthly. Hygienic conditions being what they were, floral fragrances, strong floral odours, served to camouflage or at any rate improve the polluted atmosphere where many people lived together. Roman orgies were steeped in roses - and they certainly were fragrant. The strong odours of mock orange (*Philadelphus*) and *Clematis flammula* were highly valued in Elizabethan days, and mock orange continued to serve as an odour camouflage at the small, unavoidable establishments in one wing of, or adjacent to, the main building. It certainly is a degradation from the mock orange bowers of 2-300 years ago, places for romantic - or should we say plainly erotic - adventures, to the more realistic function of later years.

These two roads led to the fragrant garden, a garden in which flowers were cultivated just as much for their fragrance as for their visual impact. There is an anecdote from Oslo at the beginning of the 19th century, told in the memories of a lady of that time. The great man of the city received a person who wanted him to give a loan, but the great man was being shaved. So he asked his servant to give Mr So-and-so a newspaper to read while he is waiting. Unfortunately, the available papers were all in English, which the person in question could not read. Confronted with this, the great man gives the next order: give Mr So-and-so that flower in the window to study. And to quote verbally: 'The servant so did, and Mr So-and-so smelt at the flower'. Our grandmothers' gardens were filled with fragrance from beautiful flowers: roses, lilies, carnations, from some not so striking ones, like wallflowers, and from some that made no pretence of possessing any visual beauty, like the mignonette. The first group still maintains its position; the wallflowers are on their way out (there are more wallflowers in the UK than in the rest of the world); and the third group, the homely ones, are all-but gone. In Norwegian poetry there is a famous poem by the dying poet singing to his wallflower: 'before your flowers have faded, I shall be the matter from which all is created'. It is known and quoted by everybody, but there is hardly any one left who knows what a wallflower is apart from custodians of the past. The mignonette has passed into complete oblivion, as have the tube rose and many others of the fragrant cohort. And when did you last meet that formerly omnipresent romantic dandy hero with a gardenia in his buttonhole? Even the rose has been endangered, not to lose favour but to lose its fragrance. Only protests from rose lovers have forced the breeders to stage a fragrance comeback.

There are many interesting fates, or misfates, among former scented favourites, like *Mimulus moschatus,* which suddenly lost its scent some time at the beginning of the century. Or that old ladies' favourite *Hesperis matronalis fl. pl. alb.* which as late as in 1939 I saw bordering long beds at a Reichsgartenschau, being told that it grew in every cottage garden in the Schwarzwald. Two years ago, when I wanted it for the reconstruction of an 18th century mansion garden, it proved impossible to obtain. The only nursery we found cataloguing it, proved not to stock it. But we had some responses saying that if we found it, would we please tell them. Apparently, in less than 50 years this plant, once so common, has virtually disappeared.

Some 200 years ago and more, people, even of high social status, loved playing pranks on each other. To our mind these pranks were often childish and of doubtful taste, to say the least, but that was the fashion of those days. Plants were involved, too, the old Scandinavian name of *Lychnis coronaria* was 'pranksters' carnation' (*fixeernellik*). If you present this to an unsuspecting person and ask him/her to smell it, the sharp paracorolla scales will prick the nose, quite distinctly. Funny? Perhaps not to us, but it had its time. Incidentally, the name appears in Norwegian seedsmen's catalogues at any rate as late as the second half of 19th century. To-day pranks like this may still survive in children's plays, but *L. coronaria* is hardly found outside the botanical gardens - if there.

I have referred to fashion. The parallelism between fashions, which we used to call styles, in gardens and in society as a whole is remarkable and fascinating. The style of the early 1800s was austere, simple, influenced by the Napoleonic wars and their aftermath. People were lean, and so was furniture: lean and rectilinear and hard with simplicity of design even if this simplicity could also mean refinement. The gardens and the individual flowers reflect the same sentiment: simple layouts, open plantings dominated by the old garden flora's European-Levantine elements.

Around 1850 a complete change-over takes place. The opulence of the industrial revolution had conquered the surroundings of urban middle class man. Furniture had become padded and fat - and so had people. We have the fat gentleman with a golden watch-chain across his fat belly, sitting in one of those fat chairs. And so was floral decoration as well: a jungle of large-leaved palms, ferns, yuccas filled every nook and cranny and flowed over from plush-covered sideboards and coffee tables and down from canterburies. A wonderful picture exists from the Grosses Hauptquartier of General von Moltke, completely filled with this floral bric-a-brac.

The late Victorian opulence gradually gave way to art noveau with its slimmer forms and love of intricate flowing patterns. How many water lilies were depicted in those days? After the First World War there was again a period of austere style - functionalism this time - and certainly many of you remember the cactus craze of those days: no more large leaves and flowing stems. And to-day: furniture has become padded again, and palms are back. Certainly they are easy to look after, but even so their return has a deeper background. The wheels have made a full turn.

In a way, roses epitomise this development. They have been in favour for the last 2,000 years at least and various centifoliate types have been known almost as long - certainly not always what we would call cv. *centifolia* and at any rate some of them were rather primitive. Both before end even more so after the introduction of the east Asiatic roses about 200 years ago rose breeders produced more strains, more and more double and more and more fragrant. Do you remember those walls covered by Gloire de Dijon with its yellow, small cabbage-head size flowers filling the garden with fragrance? And then to many of us out of the blue, around 1920 came Madame Heriot. It was simply not the same flower. Do you remember it? That exceptional orange colour and the most elegant, slender bud any rose ever had. Well, that slenderness did not conceal very much - the flower was almost empty inside and had no fragrance. But for the following decade Madame and her children were very popular: the rose answer to the cacti of the early functionalistic period. However, in the long run she was too meagre and was soon followed by fatter but weakly scented strains.

Many old garden plants have now been crowded out of European homes and gardens, partly because their cultivation has become too cumbersome, partly because new species have been introduced from exotic regions, above all because new and 'improved' strains of the old favourites have replaced them, not always for the better. Fashion takes its own unpredictable ways.

A really horrible example of lack of understanding of what happens if one neglects these aspects is provided by the park at Versailles, which the French have restored back to Le Nôtre's original concept. Beautifully done, the lakes, the vistas, the *bosquets* are all there, and so are the parterres on the upper terraces, the only place in the park where floriculture comes in. And what do we find there? Not one of the plants in those parterres was known in European horticulture at the time of Louis XIV, and most of them were cultivars created in this century: hybrid Begonias, *Achimenes, Tagetes,* and *Senecio cineraria,* not used in gardens at that time. The gaudiness of those beds matches the complete lack of understanding exhibited by the choice of plant material and the understanding of what a garden looked like at that time.

Habent fata sua. At the hands of man flowers have come, ascended to high favour and waned again. Partly man's habits have imperilled their continued existence. Partly man has quite simply rejected them, thrown them away, out of sight and often out of existence. *Habent fata sua flores.* Let us enjoy them while they are here - tomorrow they may be gone.

Gardening in the Age of Chaucer

JOHN H. HARVEY

32 Christchurch Street East, Frome, Somerset, BA11 1QH

Summary

We have no surviving gardens from the time of Chaucer but from documentary sources and from archaeology we can build up a picture of mediaeval horticulture. Chaucer himself was scientist as well as poet and wrote of his own habit of staring at the ground and 'walking in the grene'; and was himself a gardener. His knowledge of trees and plants was extensive: altogether he mentioned 159 species, 29 of them imported spices and the like. His total of 130 growing in England: trees, fruit and vegetables, herbs and flowers, compares with the 252 kinds of plants grown by his contemporary, Henry Daniel, in a private botanical garden at Stepney beside London. Daniel, who later entered the Dominican Order, was an outstanding field naturalist remarkable for his interest in ecology: the different soils and aspects of each plant. He also distinguished between plants found in the wild and those which he know only in gardens. Nor did Daniel stand alone: John Arderne (fl. 1307-1377), the famous surgeon; John Bray, the King's Physician (d. 1381); and the Master John Gardener who dictated the first English gardening book, all had outstanding knowledge of plants or their cultivation. Daniel's herbal, and that by the anonymous compiler of 'Agnus Castus', give many descriptions of plants, some of which prove that herbs used in England were substitutes for plants prescribed in continental sources. Occasional coloured illustrations throw light on the development of vegetables and flowers and show that improvement under cultivation was taking place.

Introduction

We must begin by defining our terms: gardening, the art of growing plants in a garden, dug with a spade or in later times a fork, within an enclosed area, is contrasted with agriculture, where crops are grown in open land stirred with the plough. The plants in gardens in former times were mostly herbs, *i.e.* herbaceous rather than woody, though trees and shrubs valued for medical use or for their beauty might be included. Fruit trees were commonly set by themselves in an orchard and forest trees sown, planted and cultivated in larger areas of ground, which might also serve for the preservation and hunting of deer and wild animals. The age of the poet Chaucer spans the reigns of two kings, Edward III and his grandson Richard II, and closes in AD 1400. It is thus a part of the later Middle Ages, between the fall of the Roman Empire and the onset of the Renaissance, which in England occurred near to 1500. Here, our concern begins some nine centuries after the departure of the Legions, and ends 100 years before the close of the whole mediaeval period.

As compared with the history of horticulture since 1500, what we are considering is obscure (Harvey, 1981). In the first place, there are no surviving gardens – not only from Chaucer's time, but from the whole epoch of 1100 years. A few stone walls (Fig. 1) are practically all that exists in any direct way concerned with gardening (Harvey, 1981, Fig. 78; Harvey, 1992). To build up some sort of picture it is necessary to rely almost entirely on indirect evidence: from documentary sources, including literature such as the works of Chaucer (Ellacombe, 1911; Harvey, 1981; pp. 94-95 & 122; Ruggiers *et al.*, in preparation); and from modern archaeology, which at best goes back only some 50 years (Taylor, 1983; Brown, 1991; Lambert, 1991). The evidence is scanty and has to be supplemented by reference to foreign sources, notably information as to contemporary gardens in France, the Netherlands, and western Germany. In spite of some differences, especially in regard to the species of plants grown, the evidence from this large area of north-western Europe is valid for our purpose (Fischer, 1929, pp. 257-289).

Fig. 1. Westminster Abbey: wall of the Infirmarer's Garden, *c.* 1375. Photograph copyright John H. Harvey.

Gardening and Chaucer

There is in fact a close linkage between English gardening and the poet Geoffrey Chaucer. He exemplified the universality of the educated man of his time. In spite of the fame of his remarkable literary output, he was also a scientist (Harrison, 1968). More to our purpose, he tells us of his habit of staring at the ground and of his love of 'walking in the grene' (Chaucer, *Prologue to Sir Thopas,* lines 1886-7; *Prologue to the Legend of Good Women,* line 50); there is not much doubt that he was himself an amateur botanist, and he was certainly a gardener, for he mentions 'a litel herber that I have, that benched was on turves fresshe y-grave' (Chaucer, *The Legend of Good Women,* lines 203-4). Herber, the Latin *herbarium,* meant both a garden of herbs for medical use or for cookery, and also a pleasure-garden where plants were chosen for their beauty and interest, commonly surrounded by benches covered with turf.

Without counting 29 imported vegetable products, Chaucer in his surviving works mentions 47 trees and shrubs, 34 fruits and vegetables, and 49 herbs, a total of 130 species. In some instances Chaucer's re-working of the plants listed in the French *Roman de la Rose* has importance in its own right. For example, in his account of 'many hoomly trees' he includes four not in his source: plane, yew, poplar and linden (Chaucer, *Knightes Tale,* line 2922). In the case of the Plane (*Platanus orientalis* L.) this adds to several other sources showing that the Eastern *chenar* had reached England by the fourteenth century. Chaucer's total of plants compares with the 252 kinds grown by his contemporary Henry Daniel, in a private botanical garden at Stepney beside London (Harvey, 1987, pp. 119, 127; Harvey, 1987; British Library, Arundel MS. 42, ff. 80v, 81v, 91v). By this time there is evidence of gardening in most parts of England and in southern Scotland (see map in Harvey, 1981, p. 55).

Henry Daniel

Daniel, who later became a Dominican friar, had spent seven years training as a physician. As a medical man with a sound knowledge of Latin he translated the *Circa instans* of Platearius, and upon the *Treatise of Urines* of Isaac the Jew based an extensive work of medical practice, dealing with anatomy, diagnosis and dietetics. This was compiled as a companion to the extensive herbal which constitutes his chief claim to fame. Daniel was an outstanding field naturalist remarkable for his interest in ecology. His many accounts of herbs native to Britain often give details of different soils and aspects, and show that he was in the habit of discussing botanical questions, both with fellow physicians and with 'good gardeners'. Referring to two different sorts of large cranesbill, one with blue flowers and one with red, presumably *Geranium pratense* L. and *G. sanguineum* L., he recorded that 'some suppose' them to be the same 'but diversity of soil and site maketh it to grow diversely'. As a clue to the problems being considered around 1380, when he was writing, such a comment appears significant (Harvey, 1987, p. 84; B.L., Arundel MS. 42, f. 10).

A skilled gardener, Daniel was continuing and extending the knowledge of horticultural method already shown by the Master John Gardener who had produced a treatise on gardening about the middle of the century (Harvey, 1985). Master John's 'Feate of Gardening' covered the culture of parsley; of onions, leeks and garlic; of worts, that is colewort or kale; and a long list of other herbs used for food or medicine. There are short accounts of the grafting of fruit trees and of viticulture, and a detailed section on the new crop, saffron (*Crocus sativus* L.). Master John's work, probably taken down from dictation and versified somewhere about 1340, was the only original text on the practice of gardening available in England in the later fourteenth and fifteenth centuries, apart from copies of the lengthy Latin text of the Italian Pietro de Crescenzi, which do not seem to have reached this country until after 1400 (Harvey, 1981, p. 76). In the meantime Henry Daniel had broken new ground with his detailed account of rosemary (*Rosmarinus officinalis* L.), written somewhere near 1375. This discussion of how to grow a rather tender evergreen shrub, based on more than thirty years of experience of it from its introduction about 1340, underlies all our later garden literature. It is particularly valuable for its understanding of such details as the need for water and for shelter, the seasonal behaviour of the plant, and the taking of cuttings (Harvey, 1972; 1974).

Evidence for Gardening in the Age of Chaucer

Graphic evidence for gardening is extremely sparse, and we have to go to continental sources for much of the detail. Our earliest piece of direct record is the plan of Canterbury Cathedral in *c.* 1160, which shows that there was a '*herbarium*' in the Infirmary Cloister (Urry, 1986). The garden, evidently of medicinal herbs, was enclosed by fences of trellis and had rows of plants running from east to west. More than two centuries later, the April Calendar of the now lost *Heures de Turin* shows a trellised garden in almost photographic detail (Fig. 2) (Durrieu, 1902, Plate IV; Harvey, 1981, p. 97 [Fig. 45] & 98). We can see that the plants were grown in long beds divided by narrow paths, and that the garden was bounded by walls where grapevines grew up trelliswork. A single fruit-tree stands in one bed. So much for the layout of utilitarian gardens.

Fig. 2. A garden of *c.* 1400, from the April Calendar of the destroyed *Turin Hours* (1902). From the collotype plate in Paul Durrieu, *Heures de Turin* (1902).

Quite a different tradition of gardening is exemplified by the evidence, documentary and archaeological, for the pleasaunce of the Abbot of Peterborough, Godfrey of Crowland, planted in 1302 (Fig. 3) (Harvey, 1981, p. 85). The garden was surrounded, on a square plan, with double moats crossed by bridges, and planted with pear trees (*Pyrus communis* L.) and the most lovely herbs. The cost was £25, worth at least £20,000 now. This was the abbot's private garden, quite apart from the functional gardens of the abbey and its Infirmarer and Hostillar. Pear trees were certainly planted for their ornamental value in flower, but also in leaf and in fruit. In 1264 King Henry III had had 100 pear saplings planted in the pleasure garden at Everswell by Woodstock, and their spectacular flowering was certainly the motive (Harvey, 1981, p. 80). Another quite notable tree, later in season than the pear, was the Elder (*Sambucus nigra* L.), commonly an untidy shrub, but capable of being trained as a standard with impressive results, as in recent years at Falaise in Normandy, where it was seen outside the Porte des Cordeliers, in full flower on 9 June 1976.

The Peterborough plan of a large square garden surrounded by double moats was followed at Kenilworth Castle by the Pleasance en Marys, created for Henry V in 1414-17, where there were summerhouses at three corners of the garden and a larger pavilion at the fourth angle (Colvin, 1963, **2**, p. 685; Harvey, 1981, pp. 106-107). At this point we must mention a rare survival from Chaucer's time, the garden wall of the Infirmary at Westminster Abbey, rebuilt in 1375, when a part of the abbey precinct was taken into the royal palace and the Jewel House built (Colvin, 1963, **1**, p. 535; Harvey, 1981, p. 112; Harvey, 1992, pp. 101-102).

To get an idea of the appearance of gardens of the high Middle Ages we may consider some modern re-creations, in particular that by Dr. Sylvia Landsberg at Winchester Castle, where 'Queen Eleanor's Garden' represents one of the small courtyard herbers of the late thirteenth century (Harvey, 1988). The plants are strictly those known to have been available at the time: Madonna Lilies (*Lilium candidum* L.) and Marigolds (*Calendula officinalis* L.); Red and White Roses (*Rosa* x *alba* L.; *R. gallica* L.) in a secret garden with turf benches; a stone water channel leading to the central fountain; and a tunnel arbour, all closely based on mediaeval illuminations and actual finds. A more humble garden of the fifteenth century has been reconstructed by Dr Landsberg in the Weald and Downland Open-Air Museum for 'Bayleaf', a wealden timber-framed house of the yeomanry, with a practical garden of vegetables and herbs (Harvey, 1988, pp. 26-27).

Reverting to such documentation as we have of actual gardens, we can see the beginnings of the level English lawn in an illumination of the game of Bowls of *c.* 1280, reflecting the laying and rolling of turf lawns in the Palace of Westminster in 1259 (Harvey, 1981, p. 5 [Fig. 7], 11). A notable view of a castle garden by an English artist 'Johannes', of about 1400, shows a King and Queen playing chess, seated on turf benches in a fenced garden with 'herbaceous borders', and the gardener pruning shrubs with a bill-hook (Harvey, 1981, plate III A). Within a few years the royal garden in Paris was portrayed in the *Très Riches Heures,* showing tunnel-arbours and domes. Another miniature in the same manuscript shows the Chateau de Dourdan with a lake comparable to that at Kenilworth, and a gloriette or pavilion with its walled garden surrounded by tunnel-arbours and with a chequered garden of squares of lawn and beds (Harvey, 1981, plate IV).

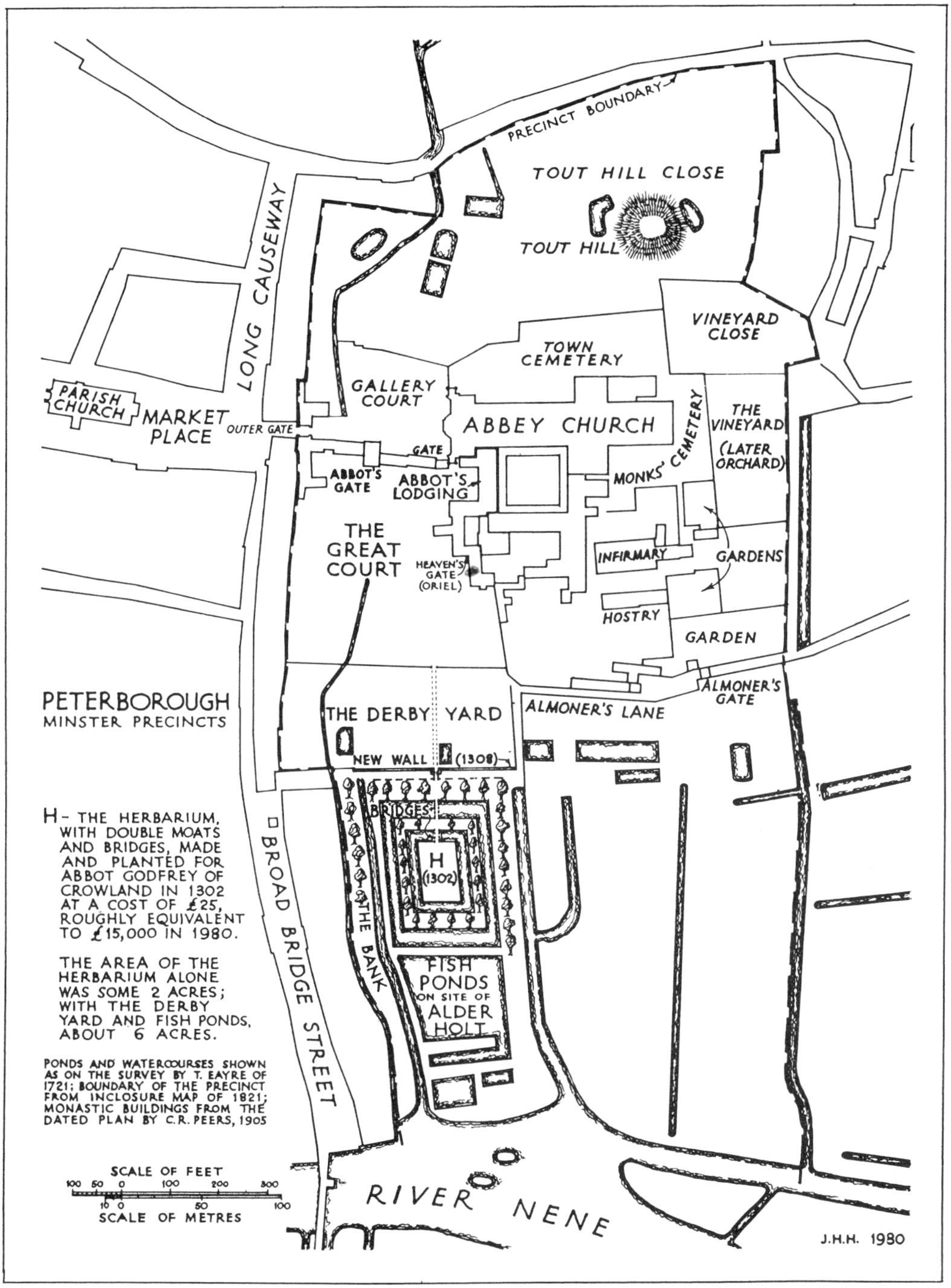

Fig. 3. Plan of Peterborough Abbey marking the monastic precinct with its gardens and vineyards, and the Abbot's Gardens, including the new 'Herbarium' formed in 1302. Drawing copyright John H. Harvey.

Other pictures from abroad display a 'Paradise Garden' of *c.* 1410-20 from the Rhineland, with many identifiable flowers (Harvey, 1981, plate V; Huxley, 1988), and Jan van Eyck's view of Maastricht of *c.* 1425 has beds of Lilies, Paeonies *(Paeonia* sp.*)* and German Iris *(Iris germanica* L.) (Crisp, 1924, Harvey, 1981, Fig. 48). By the first half of the fifteenth century the concept of the pleasure garden of ornament was stressed by a garden tapestry of trees and vases of plants in the chronicled scene of Richard II yielding the crown (Harvey, 1981, p. 5, Fig. 6). Rather later is a view of a noble garden of *c.* 1465, showing in minute detail a herber with its fencing and rails, turf benches and trees (Huxley, 1988, pp. 28-29). By way of contrast, the garden of a herbalist in a town shows square beds each devoted to a single species, a thoroughly practical approach (Huxley, 1988, pp. 24-25).

Mediaeval Physicians: Their Herbals and Herbs

Returning to Chaucer's time, Daniel's writings do not stand alone: in the same generation flourished the surgeon John Arderne, born in 1307 and still alive, in London, at seventy (Fig. 4) (Harvey, 1992, p. 100 Fig. 1, pp. 107-112; *cf.* B.L., Add. MS. 29301); John Bray, one of Edward III's physicians, who died in 1381 after compiling a long list of herbs with their names in Latin, English and French (Harvey, 1992, pp. 107-112; *cf.* B.L., Sloane MS. 282, ff. 167v-173v); and the anonymous author of a Latin herbal of *c.* 1370, later turned into English as 'Agnus Castus' (Harvey, 1987, p. 90; *cf.* B.L., Sloane MS. 2948). Arderne, Bray and this herbalist join with Daniel to throw light on the known flora of Britain, native and exotic, of their time. Comparison of their lists and descriptions provides a comprehensive picture of what was being grown in England in the reign of Richard II (Harvey, 1981, pp. 168-180; Harvey, 1992, pp. 107-112).

A few coloured illustrations throw light on botanical questions. In the Bury St Edmunds herbal of *c.* 1120 many drawings are naturalistic: the picture of a Carrot (*Daucus carota* L.) shows clearly that by the early twelfth century the root was of an orange colour, flatly contradicting the common notion that the 'red' form of the carrot did not reach Britain until much later (Oxford, Bodleian Library, MS. Bodley 130, f. 23v). The same manuscript, on a single page, shows an arbitrary nonsense for 'foxglove' (*Digitalis purpurea* L.), but a beautifully natural view of Camomile (*Anthemis nobilis* L.) (Harvey, 1981, p. 69, Fig. 38).

It is evident from many comments by Daniel that there was at the time an acute appreciation of the beauty of flowers: utilitarian considerations were not the only reason for planting a garden. On the other hand, the long tradition of the medical use of herbs fitted them into a classified strait-jacket, grouping plants by use, regardless of their natural affinities. Thus the mediaeval 'genus' *Consolida* included three plants: the great consound or comfrey, the petty consound or daisy, and the mid consound or ox-eye, respectively *Symphytum officinale* L., *Bellis perennis* L., and *Chrysanthemum leucanthemum* L. All were sovran remedies for broken bones and serious bruises, and were called in English boneworts or bruiseworts. In a similar way there was total confusion of names between the wild and garden *Solsequium:* the chicory (*Cichorium intybus* L.) and the garden solsecle, in French *souci*, the marigold (*Calendula officinalis* L.). The two plants were credited with the same virtues, as salves for sores and to cure warts, as well as being wholesome green vegetables to eat in potage.

Fig. 4. John Arderne (fl. 1307-1377) in his herb garden, from a manuscript of his works, *c.* 1400 (Glasgow University, MS. Hunter 112). Reproduced by kind permission of the University Librarian.

Study of descriptions given by Daniel and by the author of 'Agnus Castus' reveals that in some cases the plants grown in England were quite unrelated to those used in continental Europe. Thus the name *Cyclamen*, the real plant being unknown in Britain, was appropriated to the pignut (*Conopodium majus* (Gouan) Loret) [B.L., Sloane MS. 2948, ff. 6v, 25: '*Ciclamum, panis porci, malum terre* ... erthenote ... *est herba habens folia ad similitudinem Feniculi et flores albos et stipitem gracilem*' (earthnut, *Conopodium majus*, is a herb with leaves like fennel and white flowers and a slender stalk)]. Also, baldmoney (*Meum athamanticum* Jacq.) was used as a substitute for the medicinal gentian (*Gentiana lutea* L.), unrelated in the botanical sense but considered pharmaceutically identical [B.L., Add. MS. 27329, f.12: '*Allogallica, baldemonia,...genciana;* baldemoyn... it is mikil lyk wyld percile' (gentian, baldmoney... is very like wild parsley)].

Mediaeval Garden Flowers

In conclusion we have to ask how far the wild types of various species were being modified under cultivation. There is hardly any explicit evidence from before 1500 for the existence of double forms of common garden flowers other than the rose. Yet in France by the beginning of the sixteenth century there were double columbines (*Aquilegia vulgaris* L.); garden as distinct from wild daisies (*Bellis perennis* L.); garden marigolds (*Calendula officinalis* L.) as well as wild (*C. arvensis* L); double as well as single red and white carnations (*Dianthus caryophyllus* L.); double buttercups (*Ranunculus acris* L.); and large garden pansies contrasting with the wild heartsease (*Viola tricolor* L.). These are attested by the remarkable coloured drawings of Jean Bourdichon, with names both in Latin and French, in the Book of Hours of Queen Anne of Brittany which he painted in 1501-07 (Delaunay, 1859-61; the plants were identified by J. Camus, 1894).

Among the favourite garden flowers of the Middle Ages were the purple and the white forms of the rhizomatous bearded iris (*Iris germanica* L. and 'Florentina'), which are sterile clones and continue unaltered from age to age. Contrary to general supposition, they are not parents of the modern bearded flags of many colours. At the other extreme is the rose, with its enormous range of variation produced by hybridization and selection continued from remote antiquity. By the time of Chaucer the cultivated rose had already reached the state of a full double, as shown in the Wilton Diptych of *c.* 1395 (Gordon 1993, plates 3, 22, 24). This stage may only just have been reached, since the type of the perfect rose was immortalized in the Papal Golden Rose of 1309 still preserved in Paris - a semi-double of many petals but nonetheless a fertile flower (Harvey, 1981, p. 129, Fig. 73).

References

Brown, A.E. (1991). *Garden Archaeology*, ix + 198 pp. C.B.A. Research Report No. 78. London: Council for British Archaeology.

Camus, J. (1894). Les noms des plantes du Livre d'Heures d'Anne de Bretagne. *Journal de Botanique* (ed. L. Morot), **8**, 325-335, 345-352, 366-375, 396-401.

Colvin, H.M. [ed.] (1963). *The History of the King's Works*, 2 vols., xxxviii + xiv + 1139 pp. London: H.M.S.O.

Crisp, F. (1924). *Mediaeval Gardens*, xiv + 140 pp. + 225 Figs. + 314 plates. New York: Hacker Art Books, reprint 1979.

Delaunay, J. (1859-61). *Le Livre d'Heures de la Reine Anne de Bretagne*. Paris: Curmer.

Durrieu, P. (1902). *Heures de Turin*. Turin: reprint, 1967.

Ellacombe, H.N. (1911). The Flowers of Gower and Chaucer. *The Gardeners' Chronicle*, 24 June - 2 September, 1911.

Fischer, H. (1929). *Mittelalterliche Pflanzenkunde*, xii + 326 pp. (Facsimile edition, Hildesheim: Georg Olms, 1967).

Gordon, D. (1993). *Making & Meaning: The Wilton Diptych*, 96 pp. London: National Gallery Publications.

Harrison, K. (1968). Vitruvius and Acoustic Jars in England during the Middle Ages. *Transactions of the Ancient Monuments Society*, N.S. **15**, 49-58.

Harvey, J.H. (1972). Mediaeval Plantsmanship in England: the Culture of Rosemary. *Garden History* **1**(1), 14-21.

Harvey, J.H. (1974). *Early Nurserymen*, xiv + 276 pp. Chichester: Phillimore.

Harvey, J.H. (1981). *Mediaeval Gardens*, xvi + 199 pp. London: Batsford (2nd revised edition, 1990).

Harvey, J.H. (1985). The First English Garden Book. *Garden History*, **13**(2), 83-101.

Harvey, J.H. (1987). Henry Daniel: a scientific gardener of the fourteenth century. *Garden History*, **15**(2), 81-93.

Harvey, J.H. (1988). *Restoring Period Gardens*, 112 pp. Princes Risborough: Shire Publications, revised edition, 1993.

Harvey, J.H. (1992). Westminster Abbey: the Infirmarer's Garden. *Garden History*, **20**(2), 97-115.

Huxley, A. (1988). *The Painted Garden*, 160 pp. Sydney: Collins

Lambert, D. (1991). *Researching a Garden's History*, 20 pp. Reigate: Landscape Design Trust.

Ruggiers, P. & Ransom, D.J. [ed.] (in preparation). *The Chaucer Encyclopaedia*. Oklahoma: Norman.

Taylor, C. (1983). *The Archaeology of Gardens*, 72 pp. Princes Risborough: Shire Publications..

Urry, W. (1986). Canterbury, Kent. In *Local Maps and Plans from Medieval England* (ed. R.A. Skelton & P.D.A. Harvey), pp. 43-58. Oxford: Clarendon Press.

Some *Flore Pleno* Specimens gathered in the Netherlands in 1678 by John Snodgrass

J. H. DICKSON

Department of Botany, University of Glasgow, Glasgow, G12 8QQ

Summary

On folio sheets bound as a book are some 850 pressed specimens gathered in 1678 from the Leiden Botanical Garden and elsewhere in the Netherlands by the Scottish medical student John Snodgrass. There are not less than 20 *flore pleno* species from several families including Brassicaceae (*Cheiranthus, Matthiola*), Liliaceae (*Fritillaria*), Ranunculaceae (*Anemone, Aquilegia, Caltha, Hepatica, Ranunculus*) and Rosaceae (*Prunus, Rosa*).

Introduction

The little that is known about the life of Scottish medical student John Snodgrass is discussed by Dickson (1994). In the spring and early summer of 1678 Snodgrass gathered and pressed about 850 specimens mainly from the Leiden University botanical garden (Hortus Academicus Lugduno-Batavus). These are mounted on folio sheets, bound into a book (now very fragile) which is kept in Glasgow University Library. In his clear hand, Snodgrass not only named but annotated the specimens, which are arranged into trees, shrubs and herbs, with a fourth and smallest section on medicinal plants. The Snodgrass herbarium is exceeded in size by the collections of Antoni Gaymans, also made at Leiden but several to 17 years earlier (Sosef *et al.*, 1987; Heniger and Sosef, 1989) and possibly also by the later herbarium of Herman Boerhave (1668-1738), now in the Sloane herbarium (vols. 319-322) at the British Museum (Natural History), London. The many annotations, sometimes extensive, made by Snodgrass are in contrast with the very sparse comments written by Gaymans in his herbaria (Sosef *et al.*, 1987; Dickson, 1994). Often Snodgrass gives the day of collection and cites Clusius as the author of the polynomial. On occasion Snodgrass lists multiple authorities, no less than eight in the case of Yew (*Taxus baccata* L.) (Dickson, 1994).

By the end of mediaeval times, *flore pleno* (double-flowered) plants were numerous and very popular in gardens (Stearn, 1984). During the 16th to 18th centuries, herbalists and botanists catalogued these *flore pleno* plants. The most famous of the early botanists at Leiden, Charles de L'Ecluse, better known as Carolus Clusius (1601; 1611), listed many, as did Caspar Bauhin in his *Pinax Theatri Botanici* of 1623, a much-used compendium which was later reprinted (Bauhin, 1671). In *Hortus Eystettensis,* a large florilegium by Basil Besler (1613), there are magnificent illustrations, drawn at life size or greater, of very many *flore pleno* plants of species gathered by Snodgrass. Early British botanical books illustrate many *flore pleno* plants. Parkinson (1629) and Morison (1680) are good examples; Morison considered Ranunculaceae at length. Snodgrass collected many *flore pleno* plants which had been listed in the 1668 catalogue of the Leiden garden by Florens Schuyl, curator from 1667 to 1669 (Schuyl, 1668). He used Bauhin's *Pinax* and many other texts, including *Hortus Eystettensis*, which would have been available at Leiden, then of one of the greatest universities in the world.

Flore pleno and doubleness are loose terms referring to flowers that appear to have a greater than normal number of petals or other parts. The complexities and precise definitions matter little in this context; see Meyer (1966) and Reynolds & Tampion (1983). The term *flore pleno*, being a Latin ablative phrase meaning 'with full or filled flowers' is not now accepted as an adjectival epithet and has been replaced by *plenus, pleniflorus, multiplex* or a cultivar name in modern horticultural literature, notably by Bergmans (1939).

Annotated List of Snodgrass *Flore Pleno* Specimens

Anemone nemorosa L. (Wood Anemone). On folio 198: *'Ranunculus nemorosus flore albo pleno CL:'* The specimen is a shoot with leaf-like bracts and a very double flower. CL means Clusius. In *Hortus Eystettensis* Besler stated 'R [anunculus] flore albo & pleno. Rarior plante'. [= *A. nemorosa* 'Albiplena'].

Aquilegia vulgaris L. (Columbine). On folio 79: *'Aquilegia flore variegato pleno Aquilegia flore albo pleno B Aquilegia flore caeruleo pleno B:'* The first specimen has one double flower and buds, the second has two double flowers and buds and the third has four double flowers and buds and the third has four double flowers and buds (Fig. 1). B means Bauhin. Double flowered Columbines had long been familiar; Clusius (1601) has illustrations of four double *Aquilegia*. [= *A. vulgaris* 'Plena'].

Bellis perennis L. (Daisy). On folio 82: *'Bellis hortensis flore albo pleno major B May 28'*. The specimen is only one peduncle with a double capitulum with four protruding, obtuse involucral bracts. [= *B. perennis* var. *hortensis* L., with numerous forms; Brighton & Stearn, 1969].

Caltha palustris L. (Marsh Marigold). On folio 90: *'Caltha palustris flore pleno Aprilis 30'*. The specimen is a branched shoot with two expanded leaves, two flower buds and one double flower. [= *C. palustris* 'Plena'].

Cheiranthus cheiri L. (Wallflower). On an unnumbered folio (between 54 and 55): *'Leucoicum viride flore Luteo DS Leucoicum Arborescens flore Luteo pleno CL: Aprilis 10'*. The specimen is a shoot with five open double flowers and buds and 11 leaves. By DS, Snodgrass referred to Arnold Seyen who was the *Praefectus Hortus Botanicus* from 1670 to 1678, thus at the time of Snodgrass's sojourn at Leiden. Some of Snodgrass's specimens came from Seyen's private garden at Gouda, as for example Common Jujube (*Ziziphus jujuba* Miller); on folio 43 Snodgrass wrote 'Zizipha June 7 Ex horta Dom: Seyen'. Dom. is an abbreviation of the courtesy title Dominus, 'Master'.

Fritillaria pyrenaica L. (Pyrenean Fritillary). On folio 117: *'Fritillaria flore pleno DS ⊙ April: 30'* The specimen is a leafy stem with an almost upright very double flower. The meaning of the ⊙ symbol frequently used by Snodgrass is unclear; it can mean an annual plant but that is not the case in the Snodgrass herbarium because it was written beside many woody plants such as *Myrtus communis*. The *Fritillaria* is the only double flowered Monocotyledon among Snodgrass's specimens; there are no daffodils or tulips.

Hepatica nobilis Miller (Hepatica). On folio 134: *'Hepatica trifolia flore caeruleo pleno B: CL: Feb 28 ⊙ '* . The specimen is one leaf and one small, very double flower on its peduncle. *H. nobilis* had long been known in the double state. There

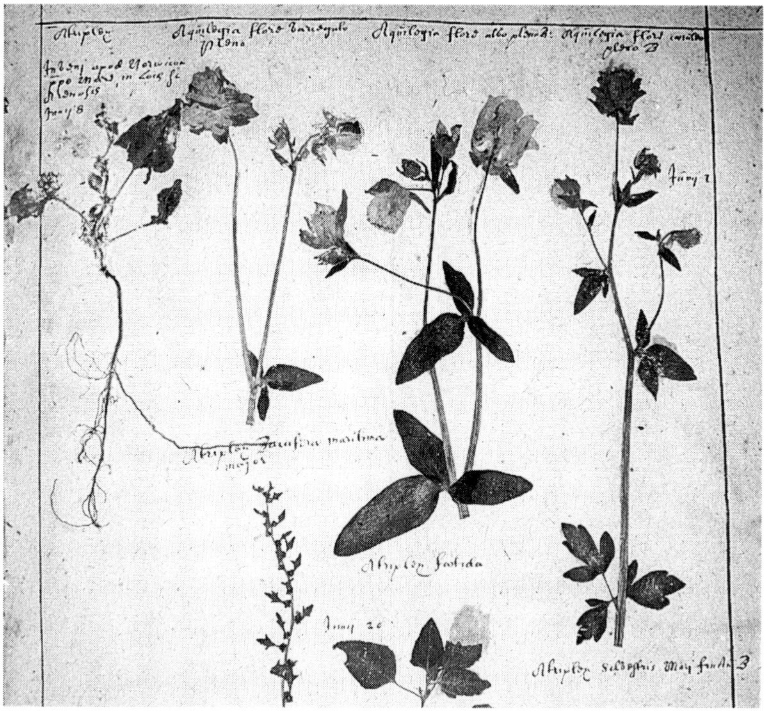

Fig. 1. Top part of folio 79 showing the three *flore pleno* specimens of *Aquilegia*. 0.4 x natural size.

is an illustration in Clusius (1601) and the plant appears on the title page of Parkinson (1629). [= *H. nobilis* 'Plena'].

Matthiola incana (L.) R. Br. (Hoary Stock). Also on the unnumbered folio (between 54 and 55): '*Leucoicum Arborescens folio Incano flore luteo holosericeo Aprilis 6*'. The specimen is a shoot with 10 leaves and six double flowers and buds.

Myrtus communis L. (Myrtle). On folio 25: '*Myrtus latifolia flore pleno CL* ⊙' His specimen is a leafy twig with no flowers (Fig. 2). In three pages devoted to *Myrtus* by Clusius (1601) there is no mention of *flore pleno* specimens. The Leiden garden grew plants from all the known world and the herbarium reflects that. However, it is no surprise that none of the *flore pleno* plants derives from the New World or southern Africa or southeast Asia. Southern Europe and the Mediterranean are well represented; *Myrtus communis*, *Fritillaria pyrenaica*, *Ranunculus bullatus* and *R. asiaticus* are examples. The timing of Snodgrass' activity meant that there would have been no flowers on the myrtle to be collected and the same would have been true for *Ranunculus bullatus*, an autumn-flowering species.

Prunus cerasus L. (Cherry). On folio 12 Snodgrass wrote '*Cerasus flore Roseo Albo CL May 10*'. His specimen is a twig with five young leaves and five umbels, each with two or three double flowers on pedicels 3-4 cm long. Snodgrass did not write *flore pleno*. However, the flowers are very double. Such *flore pleno* cherries have been known since the 17th century and are still grown. Most of Snodgrass' *flore pleno* plants are still more or less well known as cultivars now (Huxley, 1992).

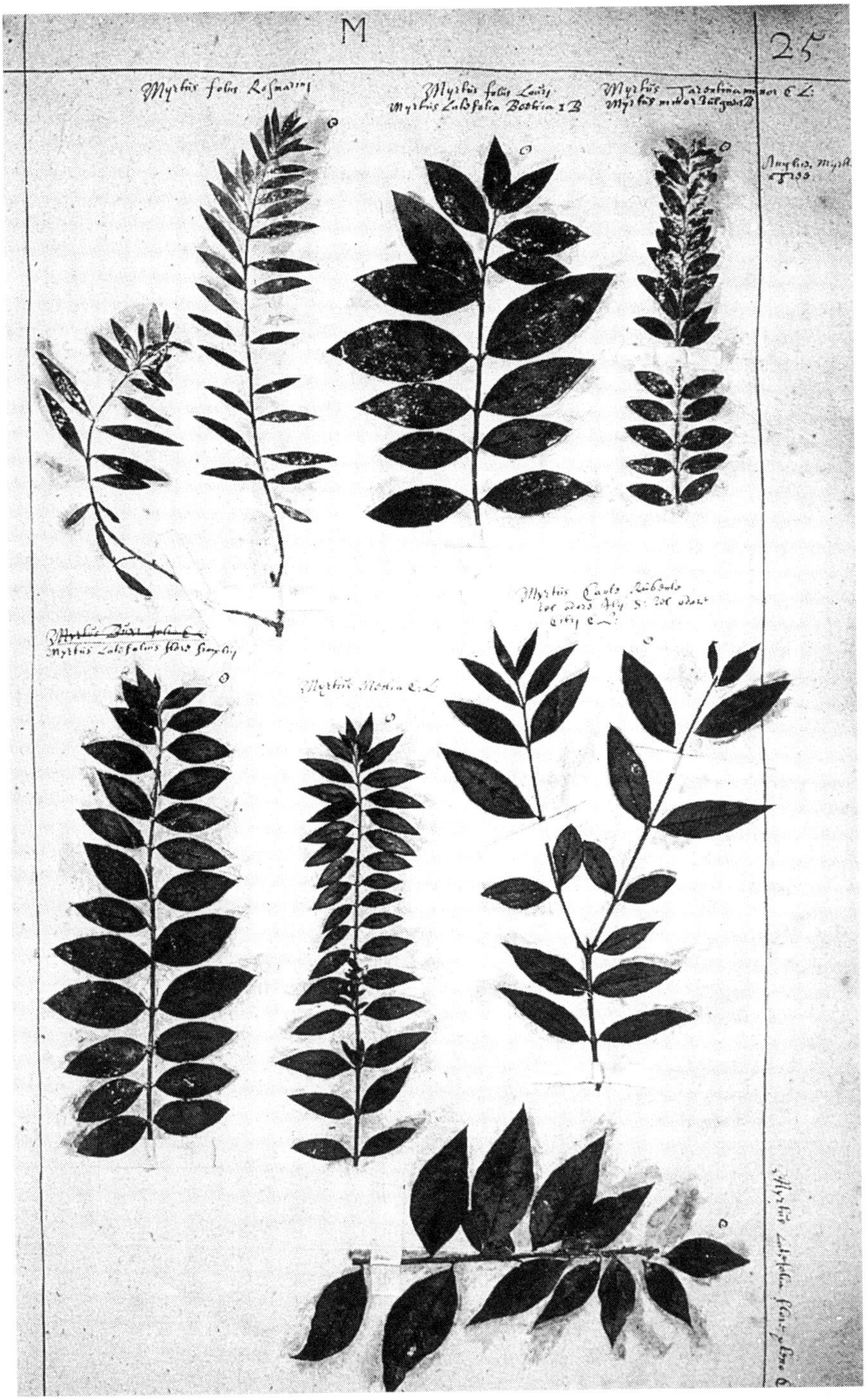

Fig. 2. Folio 25 showing seven different specimens of *Myrtus*. 0.4 x natural size. Bottom right '*Myrtus latifolia flore pleno CL*⊙'. On folio 26 are yet two more specimens of *Myrtus* with different polynomials.

Prunus persica (L.) Batsch (Peach). On folio 26: '*Malus persica flore pleno fructus confert ut plerusq(ue) plantae flores habentes plenos*' and at the margin '*est rarissima*'. The specimens are two detached flowers, labelled '*flores*', one with at least 12 petals and one with at least 14, dated '*Aprilis 12*' (Fig. 3) and a twig with nine leaves dated '*May 24*'. In Latin, Snodgrass stated that the double-flowered peach was very rare and thought it worthwhile to comment that it could bear fruit like other plants with double flowers. The *Royal Horticultural Society Dictionary* lists six *flore pleno* cultivars including 'Russell's Double' which frequently bears fruit.

Ranunculus acris L. (Meadow Buttercup). On folio 200: '*Ranunculus Hortensis inclinans flore pleno May 2*'. The specimen is one large leaf and a branched shoot with three flowers, one clearly double. On folio 201: '*Ranunculus Hortensis erectus flore pleno B Ranunculus multiplex B fuschii Ranunculus erectus Hortensis flore pleno luteo Eyst: May 10*'. The specimen is a shoot with two flowers, one clearly very double, and two leaves. Eyst means *Hortus Eystettensis*. [= *R. acris* 'Pleniflorus'].

Ranunculus aconitifolius L. (Fair Maids of France). On folio 200: '*Ranunculus folio Aconiti flore albo multiplici B Ranunculus flore albo pleno CL: C: L: May 16*'.

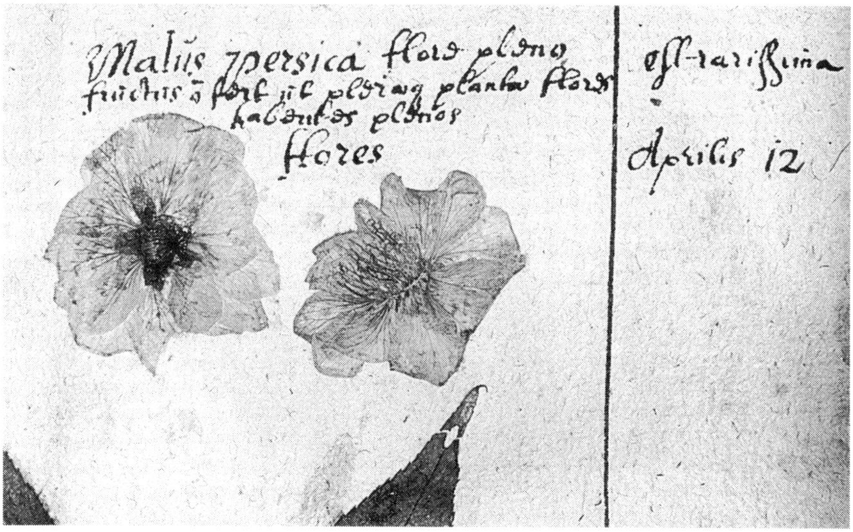

Fig. 3. Part of folio 26 showing *Prunus persica* flowers and Snodgrass' annotation. 1.1 x natural size.

The specimen is a branched shoot with four very double flowers. [= *R. aconitifolius 'Pleniflorus'*].

Ranunculus asiaticus L. (Turban Buttercup). On folio 199: '*Ranunculus flore rubro pleno. May 8*'. The specimen is a single stem with terminal double flower and three leaves.

Ranunculus bullatus L. On folio 198: '*Ranunculus autumnalis flore luteo pleno, vel Ranunculus Lusitanicus*'. The specimen is two large leaves with long petioles and no flowers.

Ranunculus ficaria L. (Lesser Celandine). On folio 90: '*Chelidonia minor flore pleno Dom: S: Aprilis 30 Scotis pilewort*'. The specimen is one leaf and one double flower on a long stalk. Bauhin noted (1671, p. 309) '*rarius pleno, Camerario monente*'. Snodgrass sometimes wrote vernacular names: Dutch, French and often English and Scottish separately. The name Pilewort is not specifically Scottish; it refers to the haemorrhoid-like root tubers, which have been found in prehistoric contexts suggestive but not conclusive of medicinal use (Dickson, 1992), and the tubers are still used in homoeopathy.

Rosa centifolia L. (Cabbage Rose). On folio 60: '*Rubus idaeus spinosus June 14*'. The specimen is a flowering stem 18.5 cm long with two leaves, uniserrate leaflets to 7 cm long, sparse, straight prickles and acicles, three tight flower buds with glandular hypanthia and pedicels and entire sepals. It may be presumed that only absent-mindedness made Snodgrass label the specimen with the formal Latin phrase name of raspberry: such lapses are very rare. The Schuyl catalogue (Schuyl, 1668) has 18 polynomials of *Rosa* and the Gaymans herbaria have ten taxa (Sosef *et al.*, 1987).

Silene behen L. On folio 153: '*Lychnis flore albo pleno June 12*'. The specimen is an inflorescence with a mass of flowers rendered indistinct by the pressing.

Trollius europaeus L. (Globe Flower). On folio 202: '*Ranunculus flore globoso pleno CL: D: CL:*'. The specimen is a single flower on a leafy shoot, with only 11 perianth segments and the other floral parts clearly differentiated. This is a normal, not double, flower. The D may refer to Rembertus Dodonaeus, author of *Crüijdeboeck* (Dodonaeus, 1554).

Viburnum opulus L. (Guelder Rose). On folio 38: '*Sambucus Aquaticus flore globoso pleno May 21*'.The specimen is a branched twig with four pairs of leaves and two inflorescences. There are no enlarged marginal flowers and stamens appear to be lacking. Snodgrass exactly followed Bauhin's polynomial (1671, p. 456). This is the commonly grown sterile cultivar of *V. opulus*, the form to which the name Guelder (or Gelder) Rose was originally applied, it having come from the Dutch province of Gelderland. The flowers are not truly double, despite Linnaeus' remark in *Hortus Cliffortianus* (1737, p. 109: 'revera flore pleno'):[= *V. opulus* 'Roseum'*]*.

Concluding comment

By dealing with only small selections of specimens, this paper and Dickson (1994) have attempted to demonstrate that the Snodgrass herbarium contains much of botanical and horticultural interest. Such is the fragility of the specimens and the loosened state of the 'book' that restoration is essential before any intensive study begins.

Acknowledgements

I am grateful to my Glasgow University colleagues Dr J. Durkan and Dr D. Weston and to Prof W. T. Stearn for their helpful comments on this paper.

References

Bauhin, C. (1671). *Pinax Theatri Botanici*, ed. 2, (i-xxvi), 518 (22) pp. Basel: Ludovi ci Regis.

Bergmans, J. (1939). *Vaste Planten en Rotsheesters*. 2nd ed., 978 pp. Tegelen: J. H. Faassen - Hekkens.

Besler, B. (1613). *Hortus Eystettensis*, 2 vols., 367 plates. Nürnberg.

Brighton, C. & Stearn, W. T. (1969). The history and cytology of double daisies (*Bellis perennis* var. *hortensis*). *Journal of the Royal Horticultural Society* **94**, 403-408.

Clusius, C. (1601). *Rariorum Plantarum Historia*, (i-iv), (v-xiv), 364, cccxlviii *pp*. Antwerp: Joannem Moretum .

Clusius, C. (1611). *Curae Posteriores Altera Appendix ad Rariorum Plantarum Historiam Caroli Clusi*. Antwerp.

Dickson, J. H. (1992). Some recent additions to the Quaternary flora of Scotland and their phytogeographical, palaeoclimatic and ethobotanical significance. *Acta Botanica Fennica* **144**, 51-58.

Dickson, J. H. (1994, in press). The dried plant collection made by Scottish student John Snodgrass at Leiden in 1678. In *Garden History Species, Forms and Varieties from Pompeii to 1800*, (ed. D. Moe, J. H. Dickson & P-M. Jorgensen).

Dodonaeus, R. (1554). *Crüijdeboeck*, (xxxix*), 418, (20) pp. Antwerp: Jan vander Loe.

Heniger, J. & Sosef, M. S. M. (1989). Antoni Gaymans (ca 1630-1680) and his herbaria. *Archives of Natural History* **16,** 147-168.

Huxley, A. (ed.) (1992). *The New Royal Horticultural Society Dictionary of Gardening*, 4 vols., 3353 pp. London: Macmillan.

Linnaeus, C. (1739). *Hortus Cliffortianus*, x, iv, 1-231, 301-501, (1, 16) pp. Amsterdam.

Meyer, M. V. (1966). Flower abnormalities. *Botanical Review* **32**, 165-195.

Morison, R. (1680). *Plantarum Historiae Universalis Oxoniensis*, vols. 2 & 3, 619 & 657 pp. Oxford: Theatro Sheldoniano.

Parkinson, J. (1629). *Paradisi in Sole Paradisus Terrestris*, (xii), 612, (16) pp. London: Humfrey Lownes & Robert Young.

Reynolds, J. & Tampion, J. (1983). *Double Flowers: a Scientific Study*, 183 pp. London: Pembridge Press, Polytechnic of Central London.

Schuyl, F. (1668). *Catalogus Plantarum Horti Academici Lugduni - Batavi*, 74 pp. Heidelbergae: Zubrodt.

Sosef, M. S. M., van Vliet, G. J. C. M., van der Meijden, R. and Scannell, M. J. P. (1987). *Catalogue of the Herbaria of Antoni Gaymans (± 1630-1680)*. Leiden: Rijksherbarium.

Stearn, W. T. (1984). The introduction of plants into the gardens of Western Europe during 2,000 years. *Australian Garden Journal* ***1984***, *Supplement*. 8 pp.

Botanical Exploration by the Academy of Turku (Finland) in the 17th Century (to the Beginning of the 18th Century)

TERTTU LEMPIÄINEN

Institute of Biology, Department of Botany, University of Turku, SF-20500 Turku, Finland

Summary

The Academy of Turku was founded in 1640 in Turku. Among the first eleven professors there was a Professor of Physics and Botany. The foundation of the Academy meant that botanical exploration became established in Finland. Elias Tillandz was the second Professor of Medicine at the Academy. He has been regarded as the first representative of the empirical natural sciences in Finland. In 1673 he published a small Latin book called *Catalogus plantarum*. This was the first flora of Finland. Tillandz's catalogue tells also about the garden plants and useful plants of Finland and the Turku district cultivated in the 17th century. After Tillandz, there was a pause of many decades in the history of Finnish botany. In the 18th century, a new chair in the Academy of Turku, an economics professorship, was founded. Its first occupant was Pehr Kalm. His best-known field trip is his journey to North America (1747-1751). His book of the journey, '*Journey to North America*', is a descriptive document of the American world at that time.

Introduction

The Academy of Turku, the first university of Finland, was founded in 1640 in Turku. Among the first eleven professors of this Academy was a Professor of Physics and Botany. The foundation of the Academy of Turku also meant that botanical exploration became established in Finland. For centuries, many young Finnish men had studied natural sciences at foreign universities, in, *e.g*, Uppsala, Tarto, Leiden, Tübingen and Wittenberg, but their knowledge bore no fruit when they came back home.

Natural sciences in the 17th century were still bound to a place. The theories of old authorities and speculation on metaphyscial problems were passed over rather quickly in Finland and exploration was concentrated on more concrete and important research topics relating to Finland. Already during the first decades of its existence the university was the centre of these interests. Botany was included among the first branches of exploration; at first it was connected to instruction in physics and belonged to the Faculty of Medicine.

The first Professor of Physics and Botany was Georgius Alanus, who came from Ahvenanmaa (Åland, SW Finland) and studied in Uppsala. He taught natural sciences for a very short time, because he was soon elected Professor of the Theological Faculty. At that time it was very typical that the teachers of the Faculty of Theology were more respected or at least were paid a higher salary than others. Thus, many capable men from other faculties often wanted to work in that faculty. This, of course, did not have a very good effect on the development of other faculties.

The successor of Alanus was a Finn, A. Thauvonius, who also left to teach theology. He did, however, teach botany for ten years. It was said of him that he appreciated personal observations made in nature and that he made field trips, but he was also inclined to resolve the problems of nature as speculatively as Alanus. After Thauvonius, the teaching of natural sciences was given to A. Thuronius, a Professor of Logic and Metaphysics. His successor, A. Petraeus, taught for twelve years until in 1682 he also went to the theological faculty.

The 17th century was not yet a flourishing time for the natural sciences, nor for medicine. As a rule the same knowledge and dogmas were taught from one generation of professors and students to the next. There was no empirically or experimentally produced new exploration and knowledge.

The number of students at the Academy of Turku rose from 200 to 1500 during the period 1620-1700 (Klinge, 1987) (Fig. 1).

Fig. 1. The building of the Academy of Turku in the 18th century (C.A. Hällström 1795/Klinge 1987).

The first Professor of Botany in Finland - Elias Tillandz, an international professor

An outstanding exception in this respect of the 17th century was the second Professor of Medicine at the Academy of Turku, Elias Tillandz, who died 300 years ago [in 1693 - *Ed.*]. Tillandz has been regarded, with good reason, as the first representative of the empirical natural sciences in Finland. The phrase 'natural science' is appropriate in this connection, because Tillandz must be remembered, first of all, for his botanical merits, although he was also, as a promoter of medicine in the 17th century, in a class of his own. He brought the exploration of nature to Finland and was a pioneer both in botany and medicine. He raised the teaching and the exploration of the young Academy of Turku to quite a new level.

In the earliest phase of botany it was important to be familiar with medicinal (herbal) plants and their medicinal effects. At the Academy of Turku, in its Faculty of Philosophy there was in the 17th century a Professor of Physics and Botany, as was mentioned above, whose duty it was to lecture in equal measure about general natural sciences and about botany. At first, in the Faculty of Medicine of the Academy there was only one professor in post. Thus, the Professor of Physics and Botany from the Faculty of Philosophy had a duty to participate also in the teaching of medicine. That was the reason why botany was mentioned in the title of the professorship (Wein, 1930; Kari, 1940; Klinge, 1987).

Tillandz was born in 1640 in Rogberga, Sweden. He was 30 years old when he became Professor of Medicine in 1670. He had recently presented his dissertation in Leiden, Holland and had studied earlier both in Turku and in Uppsala.

When Tillandz came to Turku, he had had an excellent education in medicine as well as in botany. It was necessary in the practical work of a doctor at that time to identify plant species and know their medicinal effects, as well as to be able to make medicines himself, mainly from plants. The first chemists' shops in Finland date back to 1689 in Turku and Viborg.

The history of Tillandz' family name is curious. His father, Erik Tillander, was a parson in Rogberga parish in Smooland. The son, Elias Tillander, changed his name after he had been saved from drowning to the form Til-Landz (on land). The written form in two parts he continued to use faithfully.

The first Flora of Finland - *Catalogus Plantarum*

Tillandz is known especially as a botanist and botany was his special hobby. He brought with him from Holland his own dried plant collection. He made excursions with his students in the district of Turku. In 1673 he published a small Latin book entitled '*Catalogus plantarum tam in excultis, quam in cultis locis prope Aboam superiore aestate nasci observatarum*' (A catalogue of the natural and cultivated plants which last summer were observed to grow in the district of Turku). This document, the first Finnish plant list, enumerated 496 species. Although very modest, it was the first paper on natural science which was based on original material collected from nature (Wein, 1930). It is important to note that this paper was published 60 years before Linné had classified plants and created his binomial scientific nomenclatural system. Tillandz used many different models and he was obliged to formulate Latin names for several plants he had found.

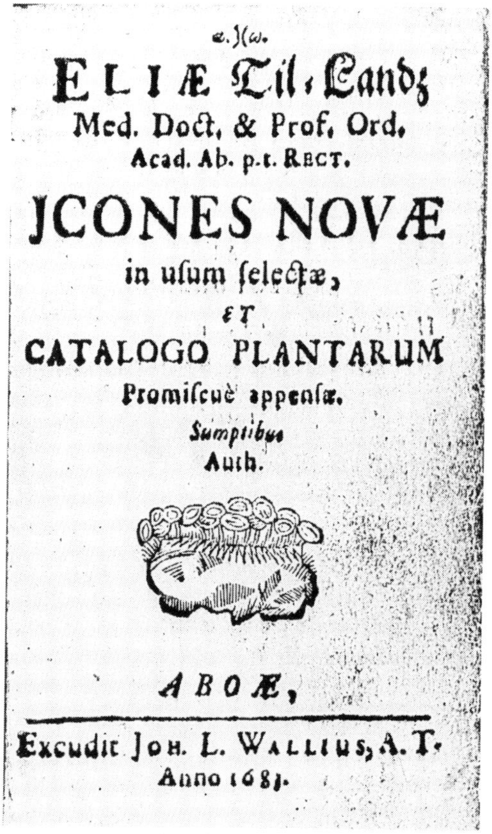

ELIÆ Til-Landz
Med. Doct. & Prof. Ord.
Acad. Ab. p.t. RECT.
JCONES NOVÆ
in usum selectæ,
ET
CATALOGO PLANTARUM
Promiscuè appensæ.
Sumptibus
Auth.

ABOÆ,
Excudit JOH. L. WALLIUS, A.T.
Anno 1683.

Fig. 2. Elias Tillandz' *Catalogus Plantarum* (1683), the second edition with the plant pictures.

Tillandz' home laboratory and the garden of medicinal plants

To help his students and others who used and were interested in plants to identify plant species, Tillandz published ten years later (1683) a picture book of plants (Fig. 2). There were whole page wood engravings of 156 plant species (Fig. 3). Some of the illustrations were original, others were copies of plant drawings published earlier. This was the first illustrated flora of Finland. There were also notes on the medicinal characteristics of plants. We can imagine how important a medium the catalogue with its illustrations was at that time for a doctor who himself was obliged to make the medicines which he prescribed for his patients. This was true of Tillandz himself, who at home had a small laboratory where he made the medicines he needed in his work.

According to the Leiden model, Tillandz also established a garden for medicinal plants, '*Hortus Medicus*', in the same area as the herb and cabbage plot of the university. *Hortus Medicus* was not only an important medium for teaching, but was also full of raw material for medicines. At home Tillandz cultivated a large number of different tropical medicinal plants. It was said to be intolerably hot there in winter (Kari, 1940).

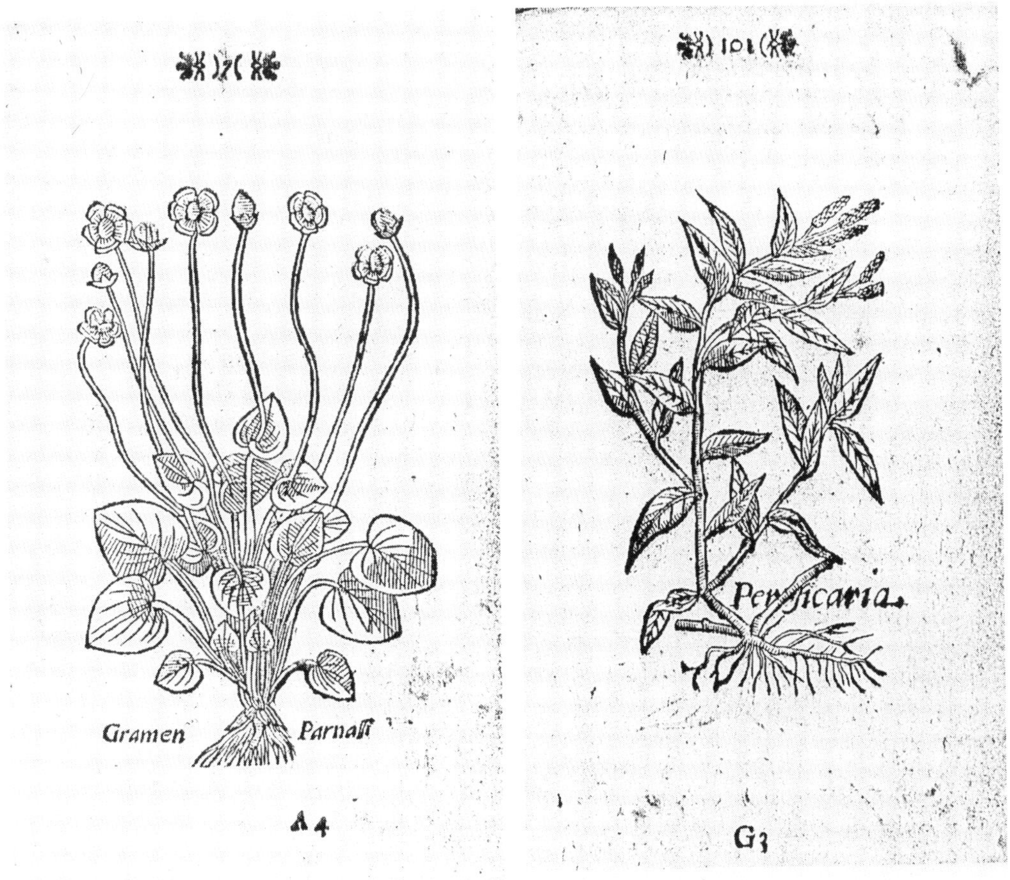

Fig. 3. Pictures from the first Flora of Finland (*Catalogus Plantarum 1863).*

Plants of Finnish gardens in the Turku neighbourhood in the 17th century

Tillandz's catalogue tells also about the garden plants and useful plants of Finland and the Turku district cultivated in the 17th century. The catalogue lists the following: apple (*Malus domestica* Borkh.), pear (*Pyrus communis* L.), plum (*Prunus domestica* L. subsp. *domestica*), cherry (*Prunus cerasus* L.), black currant (*Ribes nigrum* L,), red currant (*Ribes rubrum* L.), gooseberry (*Ribes uva-crispa* L.), pea (*Pisum sativum* L.), French bean (*Phaseolus vulgaris* L.), broad bean (*Vicia faba* L.), and different forms of cabbage (*Brassica oleracea* L.). Onion (*Allium cepa* L.), chives (*Allium schoenoprasum* L.), and leek (*Allium porrum* L.) were known, as well as many kinds of root crops: carrot (*Daucus carota* L.), beetroot (*Beta vulgaris* L.), parsnip (*Pastinaca sativa* L. var. *hortensis* Gaudin), Jerusalem artichoke (*Helianthus tuberosus* L.), black radish (*Raphanus sativus* L. *niger*), horse radish (*Armoracia rusticana* Gaertner, B.Meyer & Scherb.) and chicory (*Cichorium intybus* L.). Potato (*Solanum tuberosum* L.) was as yet unknown. Lettuce (*Lactuca sativa* L.), spinach (*Spinacia oleracea* L.) and asparagus (*Asparagus officinalis* L.) were cultivated. Many herbs were very popular: basil (*Ocimum basilicum* L.), chervil (*Anthriscus cerefolium* (L.) Hoffm.), summer savory (*Satureja hortensis* L), mint (*Mentha* L. spp.), coriander (*Coriandrum sativum* L.), anise (*Pimpinella anisum* L.) and fennel (*Foeniculum vulgare* Miller). Planted park trees included elder (*Sambucus nigra* L.), horse chestnut (*Aesculus hippocastanum* L.) and decorative shrubs were: barberry (*Berberis vulgaris* L.), and many rose species. Ornamental plants were hyacinth (*Hyacinthus orientalis* L.), tulips, narcissus, crocus, peony, columbine (*Aquilegia* L.), poppy (*Papaver* L.) and marigold (*Calendula officinalis* L.). The range of useful plants cultivated in the Turku region in the 17th century was much more varied than has been discovered by modern palaeoethnobotanical methods from the archaeological excavations in Turku (1991; Lempiäinen, in press).

Carl von Linné honoured the first botanist of Finland by giving his name to a South American genus, in the form of *Tillandsia*. The number of species of this genus of Bromeliaceae is about 400 (Mabberley, 1987). Many of them are cultivated as decorative or house plants all over the world. Thus, a curious name — the invention of a professor from Turku — lives on as a plant name in gardens, flower shops and especially among researches on tropical plants.

Tillandz as a doctor

Tillandz's work as a doctor was remarkable in Turku, because he was the only doctor in Finland who had had a modern education in medicine (Kari, 1940; Klinge, 1987). A central and important aspect of the education given by Tillandz was human anatomy. At first he had a room at home for anatomical operations in order to teach comparative anatomy, using animal corpses. In 1686, Tillandz carried out at the university a post mortem of a human body. This was a remarkable event in Turku. The rector of the university, E. Svenonius, had invited prominent people of the town and friends of science to this event. An entrance fee was collected from the guests in order to cover the costs of the post mortem (Fig. 4).

After the death of Tillandz, they found among his property a human skeleton which had been put together from several pieces. Obviously this was the object of great fear and curiosity that they had tried to peep at through the windows of his house. He left the skeleton to the university as an anatomical specimen.

Tillandz's activities for 23 years as Professor of Medicine were very important for the university and for the town of Turku. After Tillandz, there was a pause of many decades in the history of Finnish botany. The Great Nordic War interrupted the activities of the Academy of Turku (Klinge, 1987).

The time after Tillandz

In 1737 Johan Browallius was nominated as Professor of Natural Sciences. He was as old as Linné and a close friend of his youth. Browallius influenced many different fields and later became Bishop of Finland. He began the first flora of Finland, *Flora Fennica.* This large project was continued by Professor of Medicine, Johan Leche. Browallius was an unprejudiced, down-to-earth scientist. The utilitarianism of the future was seen in his opinions, but Tillandz had already been before his time in the 17th century with his empirical studies and observations, because empirical exploration came to the University of Finland as late as in the 18th century.

A new chair in the Academy of Turku, an economics professorship, was founded. Its first occupant was a pupil of Linné, the 31-year-old Finnish botanist,

Pehr Kalm. Kalm was later to become one of the most famous and remarkable scientists in the old Academy of Turku and he was internationally the best-known Finnish botanist (Kari, 1940; Kerkkonen, 1959; Klinge, 1987). Kalm made extensive field trips to study and collect seeds in Finland, Sweden, Russia and the Ukraine. His best-known field trip is, however, his journey to North America in the years 1747-1751. The purpose of this journey was to find from the area useful new plants which could thrive in the climate of Sweden and Finland and give a new impact to the economy of Finland, which was in depression after the Great Nordic War, that had been lost. The initiator and main sponsor of this journey was, along with the Academy of Turku, the Swedish Royal Academy of Sciences which had been founded in Stockholm some years earlier and to which Linné belonged as a central founder member. Kalm collected an enormous number of dried plant specimens and seeds. He also made an incredible number of careful observations about all he saw, *e.g.* human customs. His book, '*Journey to North America*', has been translated into many languages and is a valuable descriptive document of the American world at that time (Fig. 5). Pehr Kalm was one of the most famous and honoured professors of the old Academy of Turku.

Kalm's journey to North America was, as a botanical field trip, very remarkable, especially for Linné, to whom Kalm brought much of his material to identify and

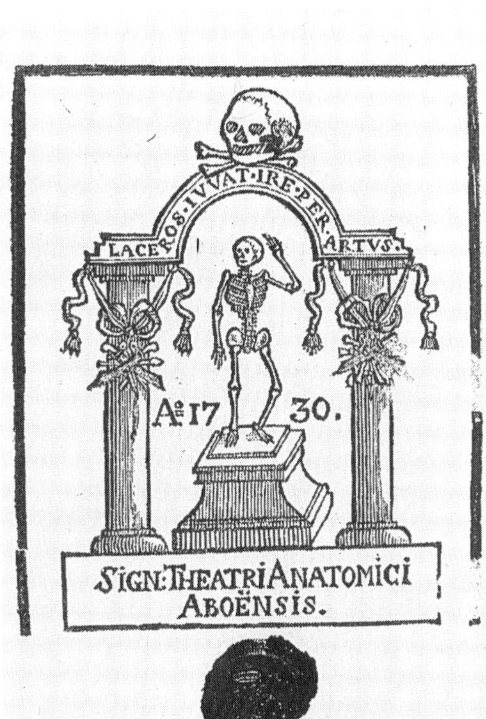

Fig. 4. A ticket for a post mortem at the Academy of Turku from the 18th century (Klinge, 1987).

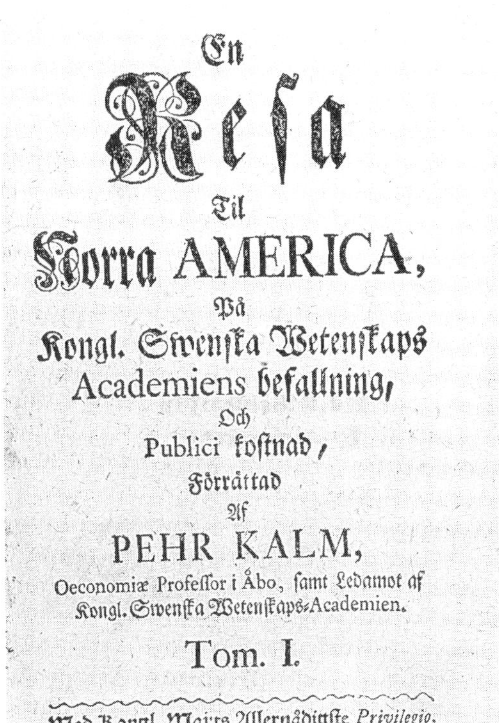
En
Resa
Til
Norra AMERICA,
På
Kongl. Swenska Wetenskaps
Academiens befallning,
Och
Publici kostnad,
Förrättad
Af
PEHR KALM,
Oeconomiæ Professor i Åbo, samt Ledamot af
Kongl. Swenska Wetenskaps-Academien.
Tom. I.
Med Kongl. Maj:ts Allernådigste *Privilegio.*
STOCKHOLM,

Fig. 5. The book of "*Peter Kalm's travels in North America*" (1753).

describe. Kalm did not want to name the plants himself but gave this work to his respected teacher. Unfortunately, only a small part of Kalm's pressed plant specimens have survived as most of them were destroyed in the fire of Turku at the beginning of the 19th century (Kerkkonen, 1959).

In 1757, in Turku, Kalm founded a botanical garden for the plants he had brought with him; this garden was to become the botanical garden of the Academy. The model was Linné's garden in Uppsala. Today all that remains of the garden is an oak by the river Aura. The experiments in cultivating the large quantity of seeds were carried out mainly outside the town, in Hirvensalo, Sipsalo and in the gardens of two parsonages nearby. Experiments were made with, *e.g.* mulberry (*Morus* L.) for the production of mulberry silkworms, and an Indian rice (*Zizania aquatica* L.). It was soon clear that most of the plants which Kalm had brought back would not thrive in the Finnish climate. Their hopes had been too great. Only a few species became established in Finland, *e.g.* Virginia creeper (*Parthenocissus quinquefolia* (L) Planchon), flowering raspberry (*Rubus odoratus* L.) and *Crataegus grayana* Eggleston. But Kalm promoted the cultivation of gardens in Finland. He supervised 146 dissertations, most in Swedish, fewer than before in Latin. He experimented in practical garden cultivation and encouraged his students to do so. In many of his dissertations there were recommendations to cultivate *e.g.* lettuce, bean, pumpkin, asparagus, artichoke, different kinds of onions, and celery (*Apium graveolens* L.) (Kari, 1940; Kerkkonen, 1957).

During the period of Utilitarianism the position of botany changed. The name of botany, *'Scientia Amabile'* ('Lovable Science'), now told only part of the truth. Botany was now important for the branch of studies leading to the use of teaching material. Its roots reached to the 17th century, to Elias Tillandz, who had, before his time, brought the useful, even necessary, empirical exploration to the then young Academy of Turku.

References

Kari, L.E. (1940). Kasvitieteen opetuksen ja kasvitieteellisen puutarhan alkuvaiheista Turun vanhassa Akatemiassa. *Varsinais-Suomen Maakuntakirja* **7**, 96-114.

Kerkkonen, M. (1959). Peter Kalm's North American Journey. *Studia Historica* **1**, 1-260.

Klinge, M. (1987). *Kuninkaallinen Turun Akatemia 1640-1808*, 740 pp. Helsinki.

Lempiäinen, T. (1991). Past occurrence of *Hyoscyamus niger* L. (Solanaceae) in Finland according to the macrofossil finds. *Annales Botanici Fennici* **28**, 261-272.

Lempiäinen, T. (in press). Macrofossil plant remains from the medieval Turku, SW Finland. *Offa.*

Mabberley, D.J. (1987). *The Plant-Book. A portable dictionary of the higher plants*, 707 pp. Cambridge: Cambridge University Press. Reprinted with corrections, 1990.

Tillandz, E. (1673). *Catalogus plantarum tam in excultis, quam in cultis locis prope Aboam superiore aestate nasci observatarum*, 15 pp. Aboae.

Tillandz, E. (1683). *Catalogus plantarum prope Aboam provenietum cum virtum recensione*, 68 + 138 pp. Aboae.

Wein, K. (1930). Elias Tillandz's "*Catalogus Plantarum*" (1683) im Lichte seiner Zeit erklärt und gewürdigt. *Vanamon julkaisuja* **11**(5), 192-411.

Collections of 18th Century Vilna Botanists in the Józef Jundziłł Herbarium

PIOTR KÖHLER

The Jagiellonian University Botanic Garden, ul. Kopernika 27, 31-501 Kraków, Poland

Summary

The herbarium of Józef Jundziłł (1794-1877), a professor of botany at Vilna University, contains plants of the 18th century Vilna botanists: Jean Emmanuel Gilibert (1741-1814) and Johann Georg Forster (1754-1794). The collection of J.G. Forster consists of 51 sheets, and there are 29 sheets of J.E. Gilibert. J.G. Forster's plants are South African species, which could have been collected during James Cook's second voyage (1772-1775). There are no data connected with the J.E. Gilibert collection.

Introduction

In the 18th century two universities existed in Poland: one in Cracow, founded in 1364, and one in Vilna (Vilnius, now the capital of present-day Lithuania), established in 1579. Towards the end of that century the Vilna University became a significant centre of botany. The history of the Vilna University Herbarium goes back to 1781, when Jean Emmanuel Gilibert (1741-1814) brought his collection of dried plants from Grodno (present Byelorussia). He also brought a considerable number of geological and zoological specimens (Sławiński, 1947).

J.E. Gilibert was a Frenchman. He studied in Montpellier, where he was conferred a doctor's degree of medicine in 1763. Before coming to Vilna, he had been a professor of botany at the College de Médecine in Lyon, and at the Royal School of Medicine in Grodno from 1776 to 1781. He lectured on natural history at the Vilna University in 1781-1783 (Mowszowicz, 1987).

The next professor (1784-1787) was Johann Georg Forster (1754-1794) who was descended from a Scottish family which had settled down near Gdańsk. Together with his father he took part in Captain James Cook's second voyage in 1772-1775 (Fedorowicz, 1987). He brought to Vilna a part of their herbarium collected during that voyage (Karavaev, 1961). After he had left Vilna, the chair of the University Department of Natural History was not filled for five years. The zoological collections were destroyed. Fortunately, the herbaria of both botanists were preserved (Sławiński, 1947; Mowszowicz, 1957).

In 1798, a Piarist, Rev. Stanisław Bonifacy Jundziłł (1761-1847), was appointed to the post of vice-professor of natural history at the University. He started gathering his herbarium in the 1780s when he began compiling his flora '*Opisanie roślin* [...]' (Jundziłł, 1791). He was abroad from 1792 to 1797 when, during disturbances connected with the Russian occupation of Poland and loss of independence by Poland, his collections vanished (Sławiński, 1947).

Józef Jundziłł and his Herbarium

Józef Jundziłł (1794-1877) was the last professor of botany before the University was closed down in 1832. He was the author of the flora of Lithuania and adjacent regions (Jundziłł, 1830), and was the first botanist to examine cryptogamic plants occurring in the vicinity of Vilna. His herbarium, handed down by his daughter, has been kept in the Herbarium of the Polish Academy of Sciences in Cracow for over 100 years. It is deposited in 40 boxes (Fig. 1) and five binders, and contains 7318 sheets (6249 being vascular plants), arranged according to Linnaeus' system. All plants, with a few exceptions, have handwritten labels; however, only 35% provide information other than the name of the species, *e.g.* locality, habitat, date or origin. In the preparation of the herbarium, J. Jundziłł used the materials from his predecessors' and students' herbaria which were housed in the Natural History Collections of the University, as well as materials from his own collection. In that way, pieces of the J.E. Gilibert and J.G. Forster herbaria became a part of his own one.

Fig. 1. Boxes 40 x 25 x 12 cm containing J. Jundziłł's herbarium.

One part of the herbarium embraces 51 sheets, whose labels, written in J. Jundziłł's handwriting, do not contain any information besides the Latin names of species, except for 'F', 'Forst', 'Forster', 'Her. Forst.', and 'H. Forster' (Fig. 2). I take these to be part of J.G. Forster's collection. Both labels and species - South African in origin - indicate it. Presumably these specimens could have been gathered during Cook's second voyage. Cook twice visited Cape Town: first in 1772, from 29 October to 22 November, and secondly in 1775, from mid March to the end of April (Cook, [ed. Barrow] 1954; Skrok, 1990). J.G. Forster writes in his diary that together with his father they started their botanical excursions on 30 October 1772. Every day they gathered a great number of plants. The house of Mr Kerste was the base for their botanical wanderings, the harvest of which filled them with apprehension, that despite their efforts they could not possibly collect, draw and preserve such a great number of specimens (Forster [transl. Ronikier], 1977). They may have completed their collections on the return journey.

There are 29 sheets in the herbarium, whose labels contain annotations by J. Jundziłł: 'Gb,' 'Gil.,' 'Her. Gil.' etc. (Fig. 3). These could indicate that these specimens came from the collections of J.E. Gilibert. This group is not homogeneous and embraces species which derive from different continents and from many climatic regions. This is probably a part of J.E. Gilibert's didactic herbarium, but, because no archival information concerning this collection has been found, it is almost impossible to ascertain its origin. There are at least three possibilities: (a) it was brought by J.E. Gilibert from France; (b) the plants were gathered in the botanical garden of the Royal School of Medicine in Grodno; or (c) they were collected after moving the botanical garden from Grodno to Vilna in 1781. Each hypothesis is equally probable.

Scientific Value of these Collections

J.G. Forster's specimens are probably the only existing part of his South African collection. Merrill (1954) described the Forsters' herbaria in detail in his paper about botanical aspects of Cook's voyages, but he did not mention anything about their specimens collected in Africa. Other authors (*e.g.* Mitrofanova, 1959) also do not write about the South African collections. I sent letters to a few herbaria which possess Forster specimens, and to scientists with expert knowledge of the Forster herbaria, in order to obtain information concerning their African collections. All replies were negative; none knew anything about Forster's South African specimens. Therefore, it is assumed that the herbarium of J. Jundziłł is the only place where the South African collection has survived.

The herbaria of J.E. Gilibert were moved from Vilna to Kiev in 1841 (Gorski, 1852). Now, specimens of J.E. Gilibert and J.G. Forster in the herbarium of J. Jundziłł are the only examples of herbaria of the first professors of botany in Poland.

Acknowledgement

The study of J. Jundziłł's herbarium was supported by the Polish Committee for Scientific Researches (project number 1 1055 9203).

Fig. 2. '*Cunonia capensis*' collected by Forster with J. Jundziłł's label '*Cunonia capensis* Forster'.

Fig. 3. '*Erica calyculata*' from the collection of J.E. Gilibert with J. Jundziłł's label '*Erica calyculata* Her. Gil.'

References

Cook, J. (1954). *Captain Cook's Voyages of Discovery* (ed. J. Barrow). *Everyman's Library* **99**, xix, 485 pp. London: J.M. Dent & Sons Ltd., New York: E.P. Dultton & Co. Inc. 1860. Last reprinted 1954.

Fedorowicz, Z. (1987). Forster Jerzy (Johann) Adam. In *Słownik biologów polskich* (ed. S. Feliksiak), pp. 164-165. Warszawa: PWN.

Forster, J. (1977). *Podróż naokoło świata* [Voyage round the world] (selected and translated by Michał Ronikier), 332 pp. Warszawa: Czytelnik.

Gorski, S. B. (1852). *Analecta ad entomographiam provinciarum occidentali-meridionalium Imperii Rossici. Fasciculus I. cum tribus tabulis coloratis aeri incisis,* xix, 214 pp. Berolini: Libraria Friderici Nicolai.

Jundziłł, J. (1830). *Opisanie roślin w Litwie, na Wołyniu, Podolu i Ukrainie dziko rosnących, iako i oswoionych podług wydania szesnastego układu roślin Linneusza* [Description of plants in Lithuania, Volhynia, Podolia and Ukraine growing wild and cultivated, according to the 16th edition of Linnaeus' system], xii, 583 pp. Wilńo: Józef Zawadzki.

Jundziłł, S.B. (1791). *Opisanie roślin w prowincyi W.X.L. naturalnie rosnących według układu Linneusza [...]* [Description of plants growing in the Great Duchy of Lithuania according to Linnaeus' system [...]], 570 pp. Wilno: W drukarni J. K. Mci y Rzeplitey, u XX Piarów.

Karavaev, M. N. (1961). G. Forster kak botanik i ego botanicheskie kollekcii v SSSR [G. Forster as a botanist and his botanical collections in the USSR]. *Trudy Instituta istorii estestvoznaniya i tekhniki AN SSSR* **36**(8), 176-201.

Merrill, E.D. (1954). The botany of Cook's voyages and its unexpected significance in relation to anthropology, biogeography and history. *Chronica Botanica* **14**(5/6), i-iv, 161-384, tab. 80-93.

Mitrofanova, N.S. (1959). Analiz gerbariya Ioganna i Georga Forsterov, khraniashchegosya v Moskovskom gosudarstvennom universitete. [Analysis of the Johann and Georg Forster's herbarium housed in the Moscow State University]. *Botanicheskiy Zhurnal* **44**(1), 135-137.

Mowszowicz, J. (1957). Conspectus florae vilnensis, pars I. *Łódzkie Towarzystwo Naukowe, Wydział III,* **47**, 9-168.

Mowszowicz, J. (1987). Gilibert Jan Emanuel. In *Słownik biologów polskich* (ed. S. Feliksiak), pp. 180-181. Warszawa: PWN.

Skrok, Z. (1990). James Cook - wielki kartograf Oceanu Spokojnego [James Cook - a great cartographer of the Pacific Ocean]. In *Odkrywcy oceanów* [Explorers of oceans] (ed. Z. Skrok), pp. 192-208. Gdańsk: Wydawnictwo Morskie.

Sławiński, W. (1947). X. Stanisław Bonifacy Jundziłł profesor Historii Naturalnej Wszechnicy Wileńskiej. The Rev. Stanisław Bonifacy Jundziłł Professor of Natural History in the University of Wilno. *Annales Universitatis Mariae Curie-Skłodowska (Lublin),* sectio **E,** supplementum **I**, 1-207, tab. I-XIII.

Sir Hans Sloane's 'Collection of Vegetables and Vegetable Substances'

A.R. VICKERY

Department of Botany, The Natural History Museum, London, SW7 5BD

Summary

The collection of 'vegetables and vegetable substances' accumulated by Sir Hans Sloane (1660-1753), and now in the Department of Botany, The Natural History Museum, London, is briefly described.

Introduction

Hans Sloane was born in County Down in 1660, and died in his 93rd year in Chelsea, London, in 1753 (de Beer, 1953; Brooks, 1954). An almost fanatic collector in the tradition of Elias Ashmole, John Tradescant and William Courten, Sloane accumulated a massive collection of natural history specimens, manuscripts, coins, minerals, and other diverse objects. After his death this collection became the nucleus of the British Museum, the natural history collections of which were moved to South Kensington in 1880, remaining part of the British Museum until 1963, when the Natural History Museum became an independent institution.

Two collections of plants accumulated by Sloane are housed in the Department of Botany of The Natural History Museum: his Herbarium, consisting of 265 bound volumes containing an estimated 120,000 specimens of pressed plants, and his 'Collection of Vegetables and Vegetable Substances', the Register of which runs to 12,523 numbers.

In 1953, the bicentenary of Sloane's death, it was decided to publish a manuscript account of the Herbarium which had been prepared by James Britten, who served on the Museum's staff from 1871 to 1909, and continued his work in the Department of Botany until his death in 1924. However, on examination it was found that Britten's manuscript was less complete than had been supposed, and extensive revision was needed before its eventual publication (Dandy, 1958).

The Collection is less well documented, and tends to receive little attention. This paper makes a start at increasing awareness of the Collection's contents, which include items of interest to botanists, students of Fortean phenomena, historians, folklorists, pharmacologists, students of plant-names, and others.

Overview of the Collection

According to Britten (1958) the Collection was started in 1701 and the last entry in the three-volume register, no. 12523, is 'Wheat and beans found Herculanum 1749'. Most of the specimens in the Collection are housed in small wooden boxes with glass tops and bottoms. The sides of the boxes were covered with decorated paper sealing in the contents. Over the years many of the original boxes have disintegrated and have been replaced with glass-topped, or cardboard, boxes.

The Collection contained some objects which have been transferred to other departments in the Museum. Thus the Bryozoa have been incorporated in the collections of the Department of Zoology, and the fossil plants (mainly from coal measures) have been incorporated in the Department of Palaeontology. There is evidence that some items were lost or given away before Sloane's death, and others have, inevitably, gone missing over the years.

Some of the objects are curiosities, or of antiquarian rather than botanical interest. Item 32 is 'a piece of the coffin of Humphry Duke of Glocester'. Item 5150 is 'A piece of Capt. Elliots canow, the bark of a tree, escaped from Algiers in the same, now placed in Greenwich hospital ... Cotton tree canoe in w[hi]ch he escaped from Petit guavas and gave notice of the French design to invade Jamaica whereby the taking of the Island was prevented'. Less colourful is the history of item 1754, 'the vertebre of an ox through the hollow of which passed an oak twigg which grew above and below it swelling'. This item is of interest in that the Register records how much Sloane paid for it - the surprisingly large sum of 10 shillings.

Contributors to the Collection include some of the people who also contributed to the Herbarium. Thus there is an extensive series of seeds from the Duchess of Beaufort's garden at Badminton. Sloane received some dried plants from the Duchess during her lifetime, but her personal herbarium, consisting of 11 volumes, was acquired by Sloane, and incorporated in his Herbarium, after her death in 1714. Most of the Duchess's items in the Collection which have a date are dated 1712 to 1714, thus it appears that these came to Sloane after her death. Included amongst the Duchess's seeds are 32 varieties of apples, and 46 varieties of pears. Thus the list in the Register provides a valuable record of at least some of the fruit trees cultivated by the Duchess.

The Collection contains material from all parts of the world, except Australasia, and it is apparent that Sloane kept a close eye on London docks for any interesting material which might turn up there, brought in by seafarers, or dug up by labourers working on extensions to the docks.

Contributors of North American material include the Reverend Cotton Mather of Boston, who sent, amongst other things, item 5478, 'Cotton? said to fall w[it]h snow from the clouds & found covering a field in New England'. Dr Richard Middleton Massey sent a parcel of seeds containing seeds of 32 species of plants from Philadelphia. These now form numbers 10945-10977 in the Register, and are followed by a list of 60 seeds given to Sloane by Mark Catesby on 25 March 1737, having been sent from Gloucester County, Virginia, by John Clayton.

Between 1687 and 1689 Sloane visited Jamaica, during which time he collected the plants which fill the first seven volumes of his Herbarium. In later life Sloane retained an interest in the Caribbean Islands, from which he acquired many specimens. For example, Henry Barham, 'an ingenious Physician in Jamaica', sent a series of items, including 7671, 'the seeds of a tree growing near the Ferry in Jamaica', and 7672, 'an aromatic bark'. From Mark Catesby came material collected in the Bahamas, including 8412, 'the fruit of the balsam tree'.

South America is less well represented. Item 6755 is 'the ordinary Paraguay tea. From Mr Josson who bought it in Buenos aires'. There are also items from the Strait of Magellan, and others which apparently came from Mexico, Panama, Surinam and Brazil.

Africa is represented mainly by specimens from Guinea, such as 7627, 'Trasi? From Guinea given my by Dr Horseman who had it off a Captain's widow. He dyed on the voyage and so could not tell what use he [?designed] to make of forty p[oun]ds of them'. Item 8121 is '[?Numum] is an herb that grows common at this place (Guinea) and most others along the Goldcoast and is made use of by the women with child to nourish the womb and outwardly applied for the pain of the head or pluritic pains. From the Duke of Chandos'. There are also items from the Cape of Good Hope, including 1607, 'a fungus' given by Dr Stuart, and an extensive series of seeds (items 6892 to 6943) from a Mr Bradbury, *via* the Duchess of Beaufort. From Egypt came 3804, 'the fruit of the lilly called faba Aegyptiaca the seeds of which are eaten there for sustenance from Mr Brickstock'.

The provenance of many of the Asiatic items appears somewhat dubious due to a rather sketchy appreciation of the geography of the eastern parts of the continent. Patrick Adair, a naval surgeon, supplied many items 'from the Indian Druggists in the East Indies', including 3142, 'long cyperus root ... called, Mut'. Unfortunately the use which was made of these items is rarely recorded. More interesting are numbers 3862 to 3926, which are Chinese drugs, apparently from the collections of the apothecary James Petiver. The Register provides details of what these were used to treat. 3917 consists of pieces of a root, described as being 'a mighty purifier of the blood'. 3922 was 'taken in powder' and used 'to remove pains in the back'. Similarly items 4429 to 4626 in the Register are a series of 'Indian druggs ... from Siam by Dr Uvedale', with details of what each of these drugs was used to treat. Other items from Asia include a good scattering of items from Sri Lanka, such as 3939, 'seeds from Zelan like Papaw seeds from Mr Robinson'.

Material from Europe includes an extensive series of seeds and fruits from a Dr Lavater, about whom nothing further can be discovered. Presumably he is not the sixteenth century Zurich physician after whom the showy Malvaceous genus *Lavatera* is named. There are also scattered items obtained from various correspondents. These include 4938, 'a small black pod from Gibraltar', and 6497, 'spurrey will be fodder for cattle in Flanders a fortnight after sowing'.

Similarly, apart from the substantial series associated with the Duchess of Beaufort, material from the British Isles includes a wide variety of objects contributed by a large number of correspondents. Apart from various petrified woods and leaves, these objects are usually mis-shapen or teratological, or are curiosities, rather than being of any real scientific interest.

Among the curiosities are seeds of beans cast ashore on the western coasts of Britain, and which Sloane, having visited Jamaica and seen the parent plants growing, was able to identify and correctly give their place of origin (Sloane, 1696). Other curiosities include 234, 'a fasciated Asparagus stalk w[hi]ch grew at Fulham given me by Mr Moore'; 2818, 'monstrous hazel nuts, three growing together from Darbyshire', and 5137, 'a mandrakes beard ... corn putt into the root of white bryony and thence sprouting?'

Some objects came to Sloane as a result of his connections with the Royal Society, of which he was the Secretary from 1693 to 1712, and President from 1727 to 1741. It is probable that the 'Tartar', 'Vegetable, Lamb', item 165 in the Collection, was acquired *via* the Society, although in its present state it bears little resemblance to the 'lamb' which Sloane exhibited at a meeting of the Society in 1698 (Sloane, 1698), or one which John Philip Breyn exhibited in 1725 (Lee, 1887). Sloane is said to have been the first person to identify the vegetable lamb as being made from a rhizome of the fern *Cibotium barometz* (L.) J. Sm., but it appears that the 'vegetable lamb', which Sloane correctly identified as being a fern rhizome, was not, in fact, a 'lamb' but a toy (possibly a toy dog) made in southern China. When these toys reached England, where the legend of the vegetable lamb had been known since the 14th century, peopled were predisposed to consider them to be lambs.

Sloane was closely involved with the Chelsea Physic Garden, which he had used as a student, and of which from 1712 he owned the freehold. There are only a few objects which can be directly associated with the Garden, including one from Chelsea's famous cedar trees: 8735, 'a branch of the cedar of Lebanon with 9 cones upon it larger than any I have from Turkey, from the physick garden at Chelsea where it had been planted ab]ou]t 50 years by Mr Wats'.

Well-known economic plants in the Collection include sugar cane (324 and 325), pineapple (326 and 327), custard apple (358, from Barbados, bought for one shilling), and 'nutmegs with mace growing round them' (462). Other reputedly useful plants mentioned in the Register include: 483, a *Strychnos* seed 'that they purify their water with in the East Indies on the coast of Coromandel or Bengale. They rubb it on the side of the earthen vessel containing the water'; 671, 'a seed called Dockonari in Persia w[hi]ch boyl'd in water makes a cooling drink; and 2296, 'a bark which the Negroes of Guinea daub themselves'.

Occasionally there are items which were used to adulterate, or replace, expensive commodities, for example, 5163, 'false coffee berries made of past and formed in a mould when the berries were dear'.

Although Sloane was active in many areas, he was primarily a physician, who received the degree of Doctor of Physic from the University of Orange in 1683, served as Physician to Christ's Hospital from 1694 until 1730, and in 1727 was appointed Physician to King George I. Thus it is not surprising that many of the objects in the Collections were considered to be of medicinal value, with their uses being recorded in the Register. Examples include: 300, 'The bark of a tree from China. It is said to cause abortion and therefore strictly prohibited by the portuguese.'; 444, 'Moxa a sort of down taken of a herb like artemisia, which they use for the cure of gout in the East Indies by burning it on the placed affected, purchased for two shillings'; 532, 'Radix Ginseng or Zinzin from China, a great restorative'; and 1508, 'A vomiting root from Virginia the dose very small from Mr Moray'.

Occasionally a medicinal use is only suspected, as in the case of 2667, 'the unknown root from Mr Bird which he conjectures to have vertues by its tast. It is bitterish to the tast'.

Ailments for which many cures are recorded include constipation (there are many mentions of 'purging'), venereal distemper, boils, the wind, coughs, and serpent stings.

Finally there are objects which are of little interest apart from what they reveal of Sloane as a collector. These are miscellaneous items of unknown origin, which many collectors would have rejected, but which Sloane boxed up and preserved. Such objects include 1512, 'bark of a tree?', and 1918, 'a pod'. However, such objects are comparatively few in number.

The Collection contains many items of interest to people working in a variety of disciplines, and it is hoped that this necessarily sketchy overview will stimulate a greater awareness of its important contents.

References

Britten, J. (1958). Sloane and his herbarium. In *The Sloane Herbarium* (ed. J.E. Dandy), pp. 15-19. London: Trustees of the British Museum.

Brooks, W.E. St J. (1954). *Sir Hans Sloane: the Great Collector and his Circle,* 234 pp. London: The Batchworth Press.

Dandy, J.E. (1958). *The Sloane Herbarium,* 246 pp. London: Trustees of the British Museum.

de Beer, G.R. (1953). *Sir Hans Sloane and the British Museum,* 192 pp. London: Oxford University Press.

Lee, H. (1887). *The Vegetable Lamb of Tartary,* xi + 112 pp. London: Sampson, Low, Marston, Searle & Rivington.

Sloane, H. (1696). An account of four sorts of strange beans frequently cast on shoar on the Orkney isles, with some conjectures about the way of their being brought thither from Jamaica, where three sorts of them grow, *Philosophical Transactions of the Royal Society* **19,** 298-300 [first two pages erroneously numbered 398, 399].

Sloane, H. (1698). A further account of the contents of the China cabinet mentioned in the last Transaction, p. 390, *Philosophical Transactions of the Royal Society* **20,** 461-462.

The Royal Botanical Institution at Amalienborg. Sources of Inspiration

PETER WAGNER

Blytsvej 6, DK 2000 Frederiksberg, Denmark

Summary

In 1752 the Danish -Norwegian government established a Royal Botanical Institution close to Frederiks Hospital, then under construction, sustained by the Privy Purse, consisting of a garden and a public library, run by a professor. The purpose of the institution was to investigate the floras in the twin kingdoms and the duchies Schleswig-Holstein by travelling through all provinces, to find plants useful in medicine, gardening and agriculture, to teach (applied) botany to scholars and non-scholars and to publish books on botany, descriptions with pictures of all plants in the realms and their uses, and to test indigenous and foreign plants for future use. German-Austrian economic theory and practice, ascribing great importance to knowledge of trades in the civil service and the population in general, furnished the background for the foundation of the institution. The fashionable interest in natural sciences is also instrumental. The didactic element was influenced by the evangelical-pietistic movement, whose members had developed a very efficient educational system, and the idea of a botanical institute connected with a hospital was probably due to Austrian inspiration. The organization of the institution was patterned after Jardin du Roi in Paris and the garden design and the travelling programme were influenced by Linnaeus. No immediate pattern for the edition of the books (*Flora Danica*) is found, but publications from Academie des Sciences might have furnished inspiration.

On October 3rd 1752 His Majesty King Frederik V of Denmark-Norway signed an ordinance founding a royal botanical library, which should be open to the public. This was the first part of the Royal Botanical Institution at Amalienborg, which consisted of a library and a botanical garden, run by a Professor Regius Botanices under the direct supervision of the Lord High Steward. The institution was closed in 1770, but one of its activities, the edition of *Icones Florae Danicae*, continued until 1887 (Wagner, 1990). But why did the king establish and sustain at his own expense by means from the Privy purse a public institution and whence came the idea of an institution with that purpose?

That kings could further botany had been demonstrated in the paragon of European states. In France, the work of the members of the Academie des Sciences on descriptions of plants in the 1660s and 1670s, resulting in Dodart's *Histoire des Plantes*, the plates of Nicholas Robert, and Tournefort's *Elemens de Botanique*, showed that it was now considered a task of the state to sustain botany (Dodart, 1676; Stroup, 1990.

The rapid development in the botanical sciences in the 18th century, as well as in agriculture and forestry, both of which in Denmark-Norway needed reorganization, made members of the government suggest the introduction of these new sciences into the twin kingdoms. As the university in Copenhagen would not cooperate, the Royal Botanical Institution was established. On earlier occasions also, the government had set up new institutions outside the university if it failed to comply. In doing so the government followed the trend in Germany and Austria, where *'Ritterakademien'* and *'Realschulen'* were established to further modern (applied) natural sciences and political science (for instance Collegium Carolinum in Braunschweig 1745 and Theresianum in Vienna 1746), a development which would be more apparent in the latter part of the century (Schmale, 1991).

Moreover, botany was fashionable. One member of the government, Count Holstein, took instruction in the Linnaean system and laid out a botanical garden; another minister, J.H.E. von Bernstorff, had a beautiful garden created, admired also outside Denmark; he and his family loved to potter in the garden (Friis, 1919). The Lord High Steward, Count Moltke, had his own collection of specimens, and the king himself not only had great interest in gardening and forestry, but even on occasion took part in garden work at the Royal summer residence at Fredensborg (Christensen, 1981).

Again a common trend in Europe was followed. Great parks and collections of rare and exotic plants had since the Renaissance been a princely sport, but towards the 1750s a change in favour of private botanical gardens, arranged to demonstrate a botanical system, took place, probably prompted by the works of Linnaeus, which made botany accessible to laymen, and by the influx of new plants from the colonies (Wagner, 1990b). Inspired by Louis de Noailles, Louis XV of France founded the botanical garden at Trianon, for which Bernard de Jussieu developed his system (Stafleu, 1963); Franz Stephan, Co-Emperor of Maria Theresia, had his own botanical garden laid out at Schoenbrunn within a year in 1753 (Jacquin, 1797), and Frederick, Prince of Wales, began the establishment of plant collections at Kew in the 1740s (King, 1985). The interest was not confined to gardens. Flower painting, flower books, and flower decorations on furniture, in interior decoration and clothes were in vogue, as were natural sciences in general. With the establishment of the Royal Botanical Institution economic theory as well as fashion were looked after.

The German-born medical doctor and botanist G.C. Oeder (1727-91), who had been called to Denmark by Bernstorff to teach botany, submitted plans for the tasks of the institute in 1752 and 1753; by examining these we may get a more intimate impression of his sources of inspiration (Oeder, 1753a).

The necessity of a botanical garden is, as it was then, evident. The herb garden of the university of Copenhagen was small, neglected and not open to the public. Oeder was then ordered to draw up a plan for a new botanical garden taking as pattern 'the arrangements of the gardens in Sweden, Cassel, Breda, Göttingen and other renowned gardens' (Moltke, 1753a), and his source is beyond doubt. In a sketch of the garden plan which Oeder sent to his teacher, Albrecht von Haller in Bern, Oeder for some details refers to the garden in Uppsala, and a comparison between the plan of this garden and his first plan for the Amalienborg garden shows the close relation between them (Wagner, 1991).

That a garden for teaching purposes was planned outside the university was not a new concept. The Jardin Royal des Plantes Medicinales in Paris, founded in 1635, had in 1718 been changed from a medical garden under the supervision of the king's head physician into a natural scientific institution, a reorganization underlined by a new name, the Jardin du Roi. Its activities as a research institution had been extended by the establishment of the Cabinet d'histoire naturelle in 1729 (Laissus, 1964), just as the project in Copenhagen was expanded with the 'natural cabinet' in 1759 following advice by Linnaeus (Wagner, 1990a). The Jardin du Roi is not mentioned by Oeder in his plans, but that was unnecessary; his protector, Bernstorff, now minister of foreign affairs, had been Danish ambassador in Paris until 1750 and was as Francophile as one could be. From the regulations of the institute signed by the king in 1761, it appears from several details that Oeder had taken for these the Jardin as a model. The conformity in purpose of the two institutions was duly recognized by the French. Having shown the plans to the *démonstrateur* during his visit to the Jardin du Roi in 1754, Oeder reported to the Lord High Steward that M. Bernard de Jussieu had honoured him, saying that his proposal was what Colbert had had in mind (Oeder, 1754). Jussieu probably referred to the Jardin as well as the Academie des Sciences; Colbert had, as protector of the Academie, lavishly funded botanical research (Stroup, 1990). The professor was expected to give instruction to enthusiasts in the garden four times a week in summer and lecture twice a week in winter. The teaching was intended for both students and laymen. This was probably due to influence from German-Austrian theory and practice in economics and political sciences (sometimes named '*cameralism*') and from pietistic influence in education.

The evangelical reform movement (pietism) had since the end of the 17th century stressed the importance of an efficient, religious school system for would-be scholars as well as for people in other trades, lessons embracing not only classical education but also new disciplines like economics, political science and mechanics. This concept was developed in Halle, at the university and in institutions created by the pietists. The concept of making knowledge and scientific experience useful/advantageous for society (*gemeinnützig*) by making it known to members of society in general, rather than keeping it within learned circles, was essential in pietism and in German political theory of the time; Oeder's plans are imbued with this concept. How close the relation is between Oeder and pietistic educational dogma is revealed in the plan for the education of prospective civil servants. This plan suggested that boys from their tenth to twentieth years should, at their parents' expense, be trained in modern technical and political science, and after a period of travel abroad be put forward for consideration by His Majesty's Service as civil servants. Details in this plan are almost identical with parts of the programme for the *mathematische, mechanische und oeconomische Realschule*, re-established in Halle in 1738 by C. Semler (1669-1740), where a botanical garden was used in the lessons (Paludan, 1885). The idea was not foreign to botanists. In a speech given in 1739 in Helmstedt on the use of botanical gardens, David Elias Heister points out that they ought to be used in teaching not only scholars but also laymen, in particular politicians and rulers; such gardens should also embellish the higher schools (Heister, 1739).

Moreover, the idea of educating laymen was not new in Denmark. In 1735 a prominent civil servant in the National College for Commerce and Economy, Count Otto Thott, had suggested a similar plan in an internal paper (Wagner, 1990a). Thott had studied in Halle on his grand tour in the 1720s, and was one of the leading civil servants. His attitude could do nothing but advance the acceptance of Oeder's plans.

The plan for the publication of *Flora Danica,* comprising not only the illustrative part but also an introduction to botany, descriptions of all the plants and their virtues and uses, was amazing. Oeder's reason for suggesting these books was that, as it was obviously impossible for a single person to collect all this information, suitably qualified members of the general public were to contribute. It was therefore necessary that as many people as possible knew the plants so that they could report their findings and make correct use of the information they were given. This connection between the institution and its users was considered very important by the government (and Oeder). The king decided in 1761 that 48 copies were to be placed in counties and dioceses for the use of those civil servants and priests 'who know best how to employ the work and in turn serve the Botanical Institute with their correspondence' (Wagner, 1990). In other words, priests and civil servants as well as gardeners, foresters and 'knowledgeable farmers' were expected to participate. These latter groups are mentioned in a proposal made for the regulations in 1760 in which experiments with foreign plants are suggested (Wagner, 1990). 'Knowledgeable farmers' are not only priests, but also gentleman farmers, whom Oeder had admired at least since his visit to England in 1754. Writing to the Lord High Steward Count Moltke, he expressed the hope that Danish noblemen would follow the example of the English and introduce modern botany in their gardening (Oeder, 1754).

The plan Oeder suggested for *Flora Danica,* with an introduction to botanical science and descriptions in the mother tongues (Danish, and later German) and with each plant depicted, shows resemblance to the works of the scientists at the Academie des Sciences mentioned above which were published in French (Tournefort's *Elemens* later being published in Latin). That Oeder dared to suggest so expensive an enterprise is probably caused by the fact that the increasing wealth in Europe in the 18th century made the publication of costly books possible, and was prompted by the craze for flowers and natural sciences in general, which dominated the fashionable world between 1730 and 1770. The public-educational element is also evident in the establishment of the botanical library as being open to amateurs as well as scholars.

The plans also included experimentation with plant drugs, a plan fitting in admirably with that for Frederiks Hospital, which was under construction, partly as a teaching hospital — yet another result of the dissatisfaction with the university on the part of those in power. It was intended that the *professor botanices* should have (and actually got) his official residence, with attached library, in one of the buildings at the hospital, and the botanical garden was laid out on adjacent ground. If one regards the combination of hospital, botanical institution and garden as a single concept together with the fact that it was financed independently of the university, through the monarch's Privy purse, the similarity to the changes Maria Theresia's physician Gerhard van Swieten introduced in Vienna in 1749 is so striking that the possibility of their having served as a model cannot be dismissed.

Among Oeder's plans were also the investigation of natural products in the colonies. To investigate the vegetation of exotic countries and colonies was not new. Already the Venetian merchants had demonstrated their connections by introducing into their gardens plants from their trade stations in the 17th century, as had the Dutch. To send scientists to collect plants for a botanical garden was not new either; Louis XIV had at the advice of Fagon sent Plumier, Feuillée and Tournefort on such expeditions (Laissus, 1985). However, the form chosen in Denmark-Norway was undoubtedly inspired by Linnaeus. Just as Linnaeus in 1746 sent his first 'apostle' to China on board a ship belonging to the Swedish East Asiatic Company (Sjörlin, 1991), so the priest Peder Middelfarth was sent to Tranquebar in India on a ship belonging to the (Danish) Asiatic Company. His instructions were written in December 1752 and the ship left Copenhagen before New Year (Larsen, undated). Oeder related to Haller that he would suggest that the king, following the example of the neighbouring Swedes, should send naturalists to the Asian and African colonies (Oeder, 1752b). Later botanical 'missionaries' were selected and educated before their departure. Among the most renowned was Johann Gerhard König (1728-1785).

It is surprising that a very important part of the duties of the *professor regius botanices* is not mentioned at all in Oeder's plan. It appears from a letter of May 1752 that Oeder was to travel in the provinces of Denmark; that in October 1753 he was to travel in Norway; and that in November he was to travel in all Danish provinces and the Duchies too (Oeder 1752a; 1753b; Moltke, 1753b), to find, describe and have illustrated the plants and record their uses. The inspiration for this is undoubtedly the travels of Linnaeus in Sweden. Linnaeus' *Ölandska och Gothländska Resa, Wästgöta-resa* and *Skånska Resa* were well known. The first books Oeder purchased for the library were those three (Catalogue of the library, 1752-59). It was also known that the travels to Öland and Gothland had been undertaken after an invitation by the Estates of the Realm, who wanted those islands investigated for economic reasons (Linnaeus, 1745); likewise the journey in Skåne took place at the king's wish. Linnaeus himself, in his speech of 1741 on the benefits of travelling in one's own country before travelling abroad (Linnaeus, 1741), had given good reasons for such undertakings. Oeder used *Flora Lapponica* extensively in Norway, and in his correspondence with Moltke from Norway Oeder again and again refers to Linnaeus. Linnaeus' argumentation for travelling and studying at home before going abroad was influenced by thoughts published by the German economic-political scientist Christian Wolff (Wolff, 1736). His views on this were known and accepted in Denmark, a fact which probably made Oeder's plans more acceptable to the government (Glamann, 1983).

The main reason for the government to establish the botanical institution was German-Austrian political-economical theory and practice, the didactic element being of German-pietistic origin. The organization of garden and institute was probably inspired by the Jardin du Roi, and the combination of garden, drug experiments and hospital by van Swieten of Vienna. The travels in the kingdoms, duchies and colonies were of Linnaean inspiration. Oeder's most original contribution was beyond any doubt the close connection with the users, which to some extent determined the form in which *Flora Danica* was edited. Works published by the members of the Academie des Sciences might have served as a kind of model.

References

Christensen, A. (1981). Frederik V's hobby: haven ved Fredensborg Slot. *Fra Kvangård til Humlekule* **11**, 35-50.

Dodart, D. (1676). *Memoires pour servir à l'histoire des plantes.* Paris.

Friis, Aa. (1919). *Bernstorfferne og Danmark. Bidrag til den danske Stats politiske og kulturelle Historie 1750-1835. II. Johann Hartvig Ernst Bernstorff i Frederik V's Konseil,* p. 295 seqq. København: Gyldendal.

Glamann, K. (1983). Et kameralistisk programskrift: Uforgribelige tanker om kommerciens tilstand og opkomst. In *Studier i dansk merkantilisme. Omkring tekster af Otto Thott* (ed. K. Glamann & E. Oxenbøll), p 44. København: Akademisk Forlag.

Heister, D. (1739). *Oratio de hortorum academicorum utilitate*, p. 21 and 26. Helmstadiae.

Jacquin, N. (1797). *Plantarum rariorum horti Caesarei Schoenbrunnensis descriptiones et icones.* I. Viennae. Introductio pp. I-II.

King, R. (1985). *Royal Kew*, pp. 60-61. London: Constable.

Laissus, Y. (1964). Le Jardin du Roi. In *Enseignement et diffusion des sciences en France au XVIIIe siècle* (ed. René Taton), pp. 287-341. Paris.

Larsen, K. (undated). *Østasiatiske personalia og realia.* File in Rigsarkivet, København.

Linnaeus, C. (1741). *Oratio, qua peregrinationum intra Patriam asseritur necessitas habita Upsaliae, in Auditorio Carolino Majori MDCCXLI Octobri XVII.* C. Haak. Lugduni Batavorum 1743.

Linnaeus, C. (1745). *Ölandska och Gothländska Resa på Riksens Högloflige Ständers Befallning förrättad Åhr 1741.* Stockholm och Upsala. Title-page.

Moltke, A. (1753a). *Letter to Oeder dated April 24th 1753.* In Bregentved Arkiv.

Moltke, A. (1753b). *Pro Memoria dated November 19th 1753.* In Bregentved Arkiv.

Oeder, G. (1752a). *Letter tó Albrecht von Haller dated May 27th 1752.* In Burgerbibliothek, Bern.

Oeder, G. (1752b). *Letter to Albrecht von Haller dated December 9th 1752.* In Burgerbibliothek, Bern.

Oeder, G. (1753a). *Plan der Stiftung vor die Botanik.* Undated but probably from the end of 1753. In Bregentved Arkiv.

Oeder, G (1753b). *Letter to Moltke dated October 26th 1753.* In Bregentved Arkiv.

Oeder, G. (1754). *Letter to Moltke dated Paris, October 8th 1754.* In Bregentved Arkiv.

Paludan, J. (1885). *Det høiere Skolevæsen i Danmark, Norge og Sverige.* Det Zetterstedtske Fond, p. 37 seqq. København: Det Zetterstedtske Fond.

Schmale, W. (1991). Die Schule in Deutschland im 18. und frühen 19. Jh. Konjunkturen, Horizonte, Mentalitäten, Probleme, Ergebnisse. In *Revolution des Wissens? Europa und seine Schulen im Zeitalter der Aufklärung (1750-1825).* (ed. W. Schmale & N.L Dodde), p. 676 Bochum: Verlag Dr. Dieter Winkler.

Sjörlin, S. (1991). Apostlarnes gärning. Vetenskap och offervilja i Linnétidevarvet. *Svenska Linnésällskapets Årsskrift. Årgang* 1990-1991, 75-89.

Stafleu, F. (1963). Adanson and the "Familles des plantes". In *Adanson. The Bicentennial of Michel Adanson's "Famille des Plantes."* Part One. (ed. G.H.M. Lawrence), pp. 140-145. Pitttsburgh: Hunt Botanical Library.

Stroup, A. (1990). *A Company of Scientists. Botany, Patronage and Community at the Seventeenth-century Parisian Royal Academy of Sciences*, pp. 65-106. Berkeley: University of California Press.

Wagner, P. (1990a). The Royal Work Flora Danica, in *Flora Danica and the Royal Danish Court.* (ed. Ole Villumsen Krog), pp. 14-42. Copenhagen: Det kongelige Sølvkammer.

Wagner, P. (1990b). Flower gardens, flower fashions and illustrated botanical works. Observations on fashions from a botanist's point of view. In *Flowers into art. Floral motifs in European painting and decorative arts.* (ed. V. Woldbye), pp. 9-23. The Hague: SDU Publishers.

Wagner, P. (1991). Den botaniske have ved Amalienborg. Oeders, Eigtveds og Thurahs projekter. *Architectura* **13,** 54-71.

Wolff, C. (1736). *Vernünftige Gedanken von dem Gesellschaftlichen Leben der Menschen*, pp. 554 seqq. Frankfurt u. Leipzig.

On the History of Vegetable Gardens in North-west Europe

A.C. ZEVEN

Department of Plant Breeding (I.v.P.), P.O. Box 386, 6700 AJ Wageningen, Netherlands

Summary
The early history of vegetable gardens is little documented. Only a few references to and depictions of mediaeval vegetable gardens are available. The possible reason is that they were too common, too vulgar, to record and to write about, and to paint or to draw. From the 16th century onwards much information is available on the history of gardens, but information on vegetable gardens is still quite limited. Many of the Flemish and Dutch paintings of the 16th to 18th centuries show vegetable stalls on markets. These paintings show us the vegetables and their variation at the time the paintings were made. More data on the history of vegetable gardens will become available as a result of archaeobotanical research.

Introduction
Quite a number of historians are attracted by the subject of the history of gardens, but they mostly restrict themselves to the well designed, well laid out gardens of wealthy individuals or families. The vegetable gardens of the common man and woman have received little attention, the main reason being that the history of vegetable gardens is not documented. As it is today, millions of vegetable gardens must have existed in the past, producing a great part of the diet.

However, this lack of knowledge may also be caused by the lack of acquaintance with vegetable gardens, either because garden historians may have grown up in a city or because they consider vegetable gardens as too common, and thus 'unimportant', to attract their attention. To obtain information on the history of vegetable gardens more attention should be paid to the limited information on vegetable gardens of the common man which is stored in paper archives. Also important as a source of information are the soil archives. So, during archaeological excavations, one should be prepared for the fact that remains of vegetable gardens and of vegetables may be found (van Haaster, 1992). Perhaps, unusually large numbers of pollen grains and macrofossils of vegetables in a given amount of soil are still present, which could indicate the early use of a piece of land as a vegetable garden. Areas of interest are those used as a vegetable garden since time immemorial, those near cottages, farms, castles and monasteries, and parcels of land now covered by buildings within and outside the former limits of a town (walled city).

In contrast to the lack of knowledge of the history of the vegetable garden, the history of the domestication of vegetables is quite well documented due to intensive research. This knowledge comes as a result of research into archaeological remains, written records, old illustrations, herbals, and the cytoplasmic and nuclear genotype (idiotype) of the wild ancestor(s), primitive forms and highly bred material. Research on DNA level provides especially revealing additional information.

In this paper I will restrict myself to written records and illustrations.

Written records

Written information should be treated with care. First, we do not know whether a single 'surviving' statement is representative of reality at that time. For instance, mediaeval medical practitioners suggested that to remain healthy one should eat only small amounts of fruits and vegetables, otherwise they cause melancholic or bad moods (*'Deghene die ghesont bliven wylt, en sal gemeynlijck niet veel fruyten noch wermoesen eten'*) (Van Haaster, 1992). However, was this advice known and applied by the common man? Was this information common knowledge written down by practitioners, or was it exchanged only between practitioners, who had got this idea into their heads? Similarly, cookery books, as today, provide recipes on how to cook, but to what extent such recipes were a reflection of commonly applied practice cannot be determined precisely. In general, many recipes could not be applied because of lack of ingredients. Furthermore, we should remember that only few people could read these books.

The written information on mediaeval vegetable gardens is very limited (Von Fischer-Benzon, 1894). One major reason is that they were probably too common and therefore of too little interest to be documented. For the Netherlands, quite an amount of late mediaeval data has been collected by Sangers (1952). The main vegetables he encountered in the archives were leeks (*Allium porrum* L.), onions (*Allium cepa* L.), garlics (*Allium sativum* L.) and peas (*Pisum sativum* L.). We do miss here coleworts (*Brassica oleracea* L.) such as perennial kale (*Brassica oleracea* var. *ramosa* (DC.) Alefeld), grown by the Romans as Tritian kale (Zeven, unpublished). Other crops were also grown, as was indicated from plant remains found at Hull, U.K. (Crackles, 1986).

Vegetable gardens must have been laid out as part of family gardens near homesteads in the country and near houses in the city (Fig. 1), and in the periphery of a city or larger village. An individual citizen may have used several vegetable gardens. The one near his house may have been cropped for vegetables. In this garden kitchen herbs, and medicinal and ornamental plants, may also have been cultivated. The gardens in the periphery of a city have been described in the Netherlands as '*koolhoven*' or '*kooltuinen*', *i.e.* coleworts gardens or cabbage gardens. These gardens were planted with leafy kales [perennial kale (*Brassica oleracea* L. var. *ramosa* (DC.) Alefeld), thousand-headed kale (*B. oleracea* L. var. *acephala* DC. subvar. *millecapitata* (Lév.) Thell.), collards (*B. oleracea* L. var. *viridis* L.), and white and red cabbages (*B. oleracea* L. var. *capitata* L. f. *alba* DC. and f. *rubra* L. (Thell.) respectively)]. These types were already mentioned as 'caboche' in South England in 1322. '*Koolhoven*', for instance, were found near Arnhem, where, in 1405, 14 '*kooltuinen*' were listed, and in 1417 an '*ortus caulinum*' was mentioned (Verkerk, 1992). In 1543 the monastery '*Het Hollandse Huis*' near Geertruidenberg, the Netherlands, ordered that some 1000 cabbage seedlings be planted in the monastery garden (Sanders, 1990). Such a number of cabbage plants would cover an area of some 400-450 m^2. More date are presented by Zeven (1993).

At a later date, more written information becomes available. For the Netherlands, most of these data have been compiled by Sangers (1952) and provide further information on the development of vegetables already grown and the cultivation of new vegetables. In the 17th century, market gardening developed

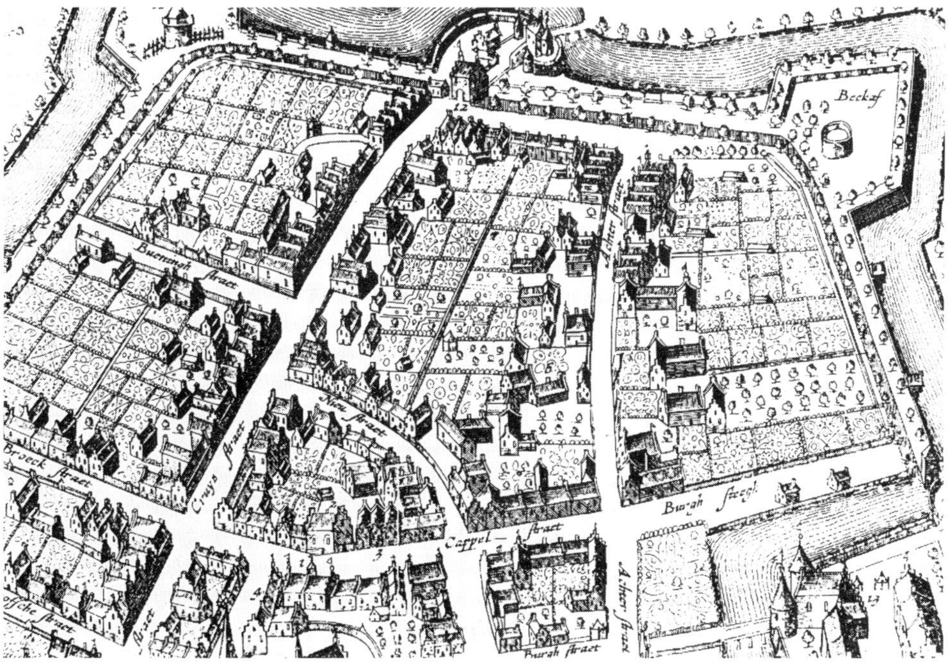

Fig. 1. A bird's eye view of a part of the city of Wageningen in 1650. A large part of the area within the wall is used for growing vegetables and other crops. Several orchards also occur.

around crowded cities such as Amsterdam, to feed their inhabitants. This development has been thoroughly described by Stone-Ferrier (1989) for Amsterdam.

Illustrations

Illustrations of plants are found on tapestries, in prayer books, and on paintings. Further, a great number of botanical drawings exist. Diehl (1954) identified the plant taxa on some 133 mediaeval and 16th century tapestries. The pea is the only vegetable found among the 115 taxa identified, whereas a cruciferous plant with yellow flowers which features in some illustrations could be a *Brassica* representative. The pea is also the only vegetable illustrated in eight mediaeval prayer books (Zeven & Stemerdink, 1986). However, I know of only one source of two illustrations of a mediaeval garden. In a copy of Pliny's *Natural History*, prepared in Venice in 1476 (Bodleian Library, Oxford), two miniatures are found. One shows a garden of palm cabbage (*Brassica oleracea* var. *acephala* DC. subvar. *palmifolia* DC., a crop still grown in Italy), and the other a garden with probably water melon (*Citrullus lanatus* (Thunb.) Matsum. & Nakai) and bottle gourd (*Lagenaria siceraria* (Mol.) Standl.) (Fig. 2). The lack of mediaeval depictions is also shown by Gothein (1926), who illustrated a contribution on mediaeval gardens with a reproduction of a painting of a backyard by Pieter de Hoogh (1629-1683)! However, the fact that there are market scenes with vegetable stalls indicates that many vegetables were grown and traded. According to Honig (1988), market scenes were described and illustrated because the 17th century citizens were proud of their local market and liked to decorate their homes with paintings showing recognizable markets. Many market scenes show a vegetable stall with its owner(s) and with one or more middle class persons in front of the stall. As most of the artists were working in West-Netherlands, *i.e.* present-day Flanders and the Dutch provinces of North and South Holland, most market scenes have been made in this area. Extremely interesting discussions of vegetable markets painted by various artists are presented by Stone-Ferrier (1989, 1992). She also described the development of horticulture in the neighbourhood of Amsterdam. Stone-Ferrier (1989) indicated that byelaws of the Amsterdam city council to regulate the trade

Fig. 2. A vegetable garden at Venice, Italy in 1476 as depicted in a copy of Pliny's *Natural History* (Bodleian Library, Oxford, U.K. Photography: LUW-fotolocatie).

of vegetables could be 'seen' on paintings. For instance, there was a byelaw that market gardeners from west and north of Amsterdam had to sell their products through women licensed by the city council. Such a woman with vegetables on her wheelbarrow is depicted by Gabriel Metsu in his painting '*The vegetable market at Amsterdam*' (dated *c.* 1661/2, Louvre, Paris).

Many species of fruits and vegetables have been depicted (Zeven, 1984; Zeven & Brandenburg, 1986; Vandommele, 1986). For several of them, for instance apple and grape, a great variation is recorded. This indicates that the various crops grown were quite variable. Furthermore, it is observed that, as in flower paintings, fruits and vegetables of various harvest seasons are often to be found together on a given painting.

Ornamental gardens

Many vegetables, kitchen herbs, medicinal plants and fetish plants have beautiful flowers. They will certainly have been used for decorative purposes. Some crops may have been developed for their decorative value and will, eventually, have been grown as ornamental plants. The desire to decorate churches may have stimulated further domestication of these plants, which may have been grown side by side with vegetables and other crops in gardens near private houses, churches, monasteries and convents.

Conclusions

The study of the early history of vegetable gardens in Europe is hampered by lack of information. This lack is caused by the fact that the actual amount of data available in paper archives and in soil archives is small, by the few illustrations of vegetable gardens and by the lack of interest of garden historians. By paying attention to the little information still preserved this study may be better founded. There are many paintings, made in the (South and North) Netherlands in the 16th and 18th centuries depicting vegetables. These paintings can be used as a source to document the vegetables grown at that time and their variation.

References

Crackles, F.E. (1986). Medieval gardens in Hull: archaeological evidence. *Garden History* **14**, 1-5.

Diehl, J.M. (1954). *Plant en dier in de middeleeuwse wandtapijten.* 49 pp. Maastricht: Publicatiereeks 7 van het Natuurhistorisch Genootschap in Limburg.

Gothein, M.L. (1926). *Geschichte der Gartenkunst.* 2nd ed. Jena.

Honig, E.A. (1988). Quirijn Gerritsz. van Brekelenkam (Zwammerdam? na 1622-ca, 1669 Leiden). In *Leidse fijnschilders. Van Gerrit Dou tot Frans van Mieris de Jonge 1630-1760.* (ed. E.J. Sluijter, M. Enklaar & P. Nieuwenhuizen), pp. 80-93, Zwolle.

Sanders, J.G.M. (1990). *Waterland als woestenij. Geschiedenis van het kartuizerklooster 'Het Hollandse Huis' bij Geertruidenberg 1336-1595,* 238 pp. Hilversum.

Sangers, W.J. (1952). *De ontwikkeling van de Nederlandse tuinbouw,* Zwolle.

Stone-Ferrier, L. (1989). Gabriel Metsu's *Vegetable market at Amsterdam:* Seventeenth-century Dutch market paintings and horticulture. *The Art Bulletin* **71**(3), 427-452.

Stone-Ferrier, L. (1992). Gabriel Metsu's *Vegetable market at Amsterdam* and its relationship to a Bredero farce. *Artibus et Historiae* no. 25 (13): 163-180.

Vandommele, H. (1986). Groenten en fruit in de Nederlanden in de zestiende eeuw. In *Joachim Beuckelaer. Het markt- en keukenstuk in de Nederlanden 1550-1650* (ed. P. Verbraeken), pp. 71-77. Gent.

Van Haaster, H. (1992). Phyto-archeologie en middeleeuws tuin-cultuur. In *Tuinen in de Middeleeuwen* (ed. R.E.V. Stuip & C. Vellekoop), pp.103-113. Hilversum.

Verkerk, C.L. (1992). *Coulissen van de macht. Een sociaal-institutionele studie betreffende de samenstelling van het bestuur van Arnhem in de middeleeuwen en een bijdrage tot de studie van stedelijke elitevorming*, 758 pp. Hilversum.

Von Fischer-Benzon, R. (1984). *Altdeutsche Gartenflora. Untersuchungen über die Nutzpflanzen des deutschen Mittelalters, ihre Wanderung und ihre Vorgeschichte im klassisschen Altertum* L., 254 pp. Kiel & Leipzig.

Zeven, A.C. (1984). Het gebruik van afbeeldingen voor de studie van gewashistorie. *Tussen Rijn en Lek* **18**(3), 25-40.

Zeven, A.C. (1993). Vegetables and vegetable gardens in North-West Europe: their history as shown by 15th to 18th century paintings. *PACT-journal* **42** (in press)

Zeven, A.C. & Brandenburg, W.A. (1986). Use of paintings from the 16th to 19th centuries to study the history of domesticated plants. *Economic Botany* **40,** 397-408.

Zeven, A.C. & Stemerdink, G.J. (1986). A cluster analysis of eight mediaeval manuscripts based on depicted plant taxa. *Travaux d'ethnobotanique et d'ethnozoologie, publ. avec le concours du CNRS.* **33,** 225-241.

Medicinal and Other Useful Plants in the Faroe Islands before AD 1800

JÓHANNES JÓHANSEN

Museum of Natural History, FR-100 Tórshavn, Faroe Islands

Summary

In this paper four authors who write about plant use in the Faroe Islands before AD 1800 are dealt with. Peder Claussøn (1545-1614) was a Norwegian priest who wrote the book *A True Description of Norway and Surrounding Islands*. He mentioned that *Potentilla erecta* (L.) Räusch. is used for tanning sheep leather. Simon Paulli (1603-1680), German professor in Copenhagen, says in his *Flora Danica* (1648) that people in the Faroes used roots of Filicales to remove stones in the kidney and bladder. Lucas Debes (1623-1675), a Danish priest who lived in the Faroes, wrote the book *A Description of the Faroe Islands and their Inhabitants*. He mentions that people cultivated *Angelica archangelica* L., which they called *kvan*, and that *Rhodiola rosea* L. gives good rose-water. Jens Christian Svabo (1746-1824), a Faroese student, wrote a voluminous report on a journey to the Faroes (1781-82) which he made for the Danish government. He gives many examples of medicinal plants and has a special chapter on plant dyeing of wool.

Introduction

The farther north one travels, the more important animal food becomes and the less important are plants. At the extreme habitats of Eskimos, the people live exclusively on animal food. However, the Faroes lie further south. They were colonized by Norwegian people in Viking times. The Vikings were farmers and consequently plants were very important to them. However, we do not find many sources before AD 1800 about plant use; in fact, only four authors give us original information.

Peder Claussøn (1545-1614)

Peder Claussøn was born in Ekernsund in southernmost Norway where he was a priest in Lista. What he writes about the Faroe Islands he received from a Faroese student in Copenhagen - Jákup Eythunsson. Peder Claussøn wrote many topographical books describing Norway, Faroes and Greenland. He also translated Snorre Sturluson's King sagas. What is cited below was written in 1592, but was not published until 1632.

Peder Claussøn writes about only one economic plant, *Potentilla erecta* (L.) Räusch.: 'Tormentil grows abundantly there (*i.e.* Faroes), which they use for tanning skins or leather because they don't have the bark of trees and it becomes good leather, which is tanned in this way; a girl or a boy should be able to dig half a barrow of tormentil a day/their swine become fat by tormentil though they don't keep many swine there'.

Tormentil has been used for tanning up to recent times. In the 1940s and 1950s it was not so uncommon to see women working on it (Fig. 1). The rhizome was used; it was crushed in a round hollow in the rock with a stone pebble. The mass was then mixed with water and smeared out on the fleece. Sometimes it was dried and used as powder.

This massive use of *Potentilla erecta* has not affected the abundance of the plant, which is one of the most widespread plants in the Faroes today.

Fig. 1. Woman working on tormentil. She crushes the rhizomes in a small hollow in the rock using a stone. Island of Koltur. Danish Folk Museum.

Simon Paulli (1603-1680)

Simon Paulli, a German born in Rostock, became Professor of Medicine and Botany at Copenhagen University. In 1648 his *Flora Danica* appeared with illustrations and text concerning about 400 Danish plants. In this large work he mentions the Faroe Islands in four lines, stating that the inhabitants use ferns as medicine. He writes: 'In the Faroes they put the root of this herb (*i.e., Filix*) into beer and use it against stones in the kidneys and bladder with good and happy success.' Who gave Simon Paulli this information is unknown. The advice to use ferns is not given by later authors writing about the Faroes.

Lucas Debes (1623-1675)

Lucas Debes was a Danish priest who in his youth came to the Faroe Islands in 1651. He wrote a book, published in 1673: *Færoæ & Færoa Reserata* [A description of the Faroe Islands and their inhabitants]. We have now for the first time a book from a man who, although Danish, had lived in the Faroes for 20 years and knew the islands personally.

As the title says the work is a general description of the Faroe Islands: about the country, the water, whales, fish, birds, fertility of the land, *i.e.* agriculture and sheep rearing. He also writes about historical events and religion. We shall consider only what he says about plants and people.

Lucas Debes writes that just as God sends illness to a country, He also put plants to grow there which could cure the illness. As there was much scurvy in the Faroes in earlier times, so there are scorbutic plants in abundance. Debes mentions *Nasturtium* (presumably *Rorippa palustris* (L.) Besser, (*Nasturtium palustre* (L.)DC.) or a species of *Cardamine* L., as *Nasturtium officinale* R.Br. does not grow in the Faroe Islands), *Cochlearia* L., *Veronica beccabunga* L. and *Rumex acetosa* L. He then discusses *Potentilla erecta* in the same way as did Peder Claussøn.

Kvan, Angelica archangelica

Debes then writes about *Angelica archangelica* L. His text runs as follows: 'There is a great abundance of *Angelica*, here called *kvan*, not only in gardens and churchyards, where it is planted with diligence, but also in the wild fields and up the high mountains. The inhabitants eagerly eat the large hollow stems, which bear no seeds. They first take the skin away, then remove the long fibres. The rest they eat for pleasure, just as people in other countries eat fruit; in bad times they even eat *Angelica* roots for food.' The word *kvan* seems to be of Norwegian origin, but its meaning is not clear (Lyttkens, 1904-15).

The oldest gardens in Scandinavia were *kvan* gardens. *Kvan* was very important in Viking times, being a major export commodity from Norway to the continent, especially France. It was mostly used for medicinal purposes, being supposed to cure almost every disease, especially the plague, but it was probably also eaten as we know it from the Faroes. Most of the old *kvan* gardens have disappeared today, but in the Faroes, and at Voss in Norway, they have persisted up to recent times. In the Faroes they were common in every village in the 1950s, but have gradually decreased in number. If they are not looked after, they are soon destroyed by sheep. *Kvan* gardens were usually very small (2-10m^2). One of these old gardens, on the small island of Koltur, is now protected and taken care of (Fig. 2).

Later authors in the Faroes tell us that *kvan* was mainly eaten by children as 'sweets' and was also served to travellers and guests. It was dipped in fresh cream and sugar and then eaten. In earlier times *kvan* was important as a valuable food supplement. From Svabo, mentioned below, we know that the Faroese people distinguished between cultivated home-*kvan* and the wild mountain-*kvan* and they regarded the mountain-*kvan* as having a much better taste.

An anatomical difference between cultivated and wild *kvan* has also evolved. The leaf stalk or petiole of cultivated *kvan* is solid and that of wild *kvan* hollow (Fægri, 1946).

One of the few Nordic words, if not the only one, which was found in the Greenland Eskimo language, when Greenland was rediscovered, was *kvan*. This can only have come from connections between the Nordic settlers and the Inuit Eskimos; it shows how important this plant was there and that the Eskimos had learned to appreciate *kvan*.

There is hardly any doubt that the settlers brought with them from Norway to the North Atlantic islands the habit of growing and using *kvan*. In support of this

Fig. 2. A *kvan* garden on the island of Koltur. This is a garden which is several hundred years old. It has been disturbed to some degree, but is now protected. Photographs by the author.

is the occurrence of *kvan* in Shetland. It does not grow in the wild, but was present in only a few gardens in South Mainland until 1973, when it became extinct. *Kvan* was also very important in Iceland (Ólafsen, 1772). In Greenland the plant has always been found around Viking ruins (Bruun, 1904), as well as growing in the wild.

However, it is equally clear that in the Faroes, Iceland and Greenland *kvan* is a native plant, growing in remote, inaccessible places. In pollen diagrams from the Faroes, pollen of Umbelliferae (including only *A. sylvestris* L. and *A. archangelica*) drops steeply at the *landnam* (Norwegian settlement) horizon. If *A. archangelica* were introduced for the first time we would expect the Umbellifer curve to rise. We know that there was a vigorous vegetation, no doubt also containing *kvan* (at any rate many Umbellifers) in the Faroes and Iceland, and that the sheep destroyed much of this original flora. *A. archangelica* only survived on the steep bird cliffs, from which people took it into their gardens. In modern times it has become a kind of hobby for some people to make a *kvan* garden. They use *kvan* to give *snaps* (an alcoholic drink) a special taste.

Rhodiola

The last plant Lucas Debes mentions is *Rhodiola rosea* L. which is extremely common in the Faroe Islands. Debes says that from *Rhodiola* he obtained the finest rose water. Both in smell and taste it is equally good as any rose water he had procured from abroad. Rose water was much used in Lucas Debes's time and later. Genuine rose water was made of rose oil, which was extracted from the oriental *Rosa damascena* Miller. It was highly appreciated both for its smell and its taste

Jens Christian Svabo (1746-1824)

Jens Christian Svabo was a native Faroese who went to Copenhagen to study natural sciences, but he never completed his studies. However, the Danish government sent him to the Faroes to report about the natural history, economy and things that could be used to improve the living standard of the islanders. The result was a long report, which remained unpublished until 1959.

Svabo (1959) mentions that *kvan* is often given not only to children but also to travellers, in the latter cases with cream and sugar added. He then gives a list of Faroese plants and sometimes also explains how the natives used them. These plants shall briefly be mentioned.

Cochlearia officinalis L. - for scurvy.

Juniperus communis L. - used to smoke meat.

Blechnum spicant (L.) Roth - used by midwives to stop bleeding at birth.

Thymus polytrichus A. Kerner ex Borbás (*T. serpyllum* auct. non L.) - is by some mixed with tobacco.

Plantago major L. - a good healing plant for wounds and external bruises.

Peltigera canina is used for skin diseases.

Equisetum L. - the small nodules on the root are eaten.

Angelica archangelica L. - Svabo mentions the difference between wild and cultivated *kvan* as is mentioned above.

Erica L. - sometimes used for smoking meat.

Tussilago farfara L. - used to heal wounds.

Ligusticum scoticum L. - used by some as a means to sleep better.

Angelica sylvestris L. - root is eaten in some places.

Juncus effusus L. - pith used as a wick in oil lamps.

Svabo has a separate chapter on dye plants, stating that the Faroese can produce five different colours when dyeing their woollen clothes. These are obtained from the following plants:

1. *Lecanora* Ach., sp. This lichen gives a beautiful red or purple colour. Modern women say it is impossible to get a good red colour. This shows the great skill in this work the women had in Svabo's time. Svabo says that the farmers' wives are as good at textile dyeing as any colour master abroad.
2. *Parmelia saxatilis* (L.) Ach. yields a very handsome and stable brown colour. Svabo states that an English sailor who overwintered in a village in the Faroes taught the Faroese how to use *Parmelia.* This would have taken place 100 years before Svabo, *i.e.* about 1680. It is very popular today.
3. *Geranium sylvaticum* L. was used together with a special kind of soil to get a completely black colour.
4. Yellow was achieved by using *Diphasiastrum alpinum* (L.) Holub (*Lycopodium alpinum* L.)
5. For green they used the alga *Ascophyllum nodosum* (L.) Le Jolis. However, Svabo writes that the knowledge about this process was almost forgotten and it was only rarely used.

References

Bruun, D. (1904). Fra de færøske Bygder. *Tidsskrift for Landøkonomi.* **24**.

Claussøn, P. (1632). *Norriges oc omliggende Øers sandfærdige Beskriffeulse,* 108 + 8 pp. Kiøbenhaffn.

Fægri, K. (1946). Angelica - trekk av dens natur- og kulturhistorie. *Naturhistorisk Tidende* **10**, 9-10.

Debes, L. (1673). *Færøæ et Færøa Reserata,* 20 + 367 pp. Kiøbenhaffn.

Lyttkens, A. (1904-15). *Svenska Växtnamn,* 1891 pp. Stockholm.

Ólafsen, Eggert (1772). *Vice-Lavmand Eggert Ólafsens og Land-Physici Biarne Povelsens Reise igienem Island, foranstaltet af Videnskarbernes Sælskab i Kiøbenhavn, og beskreven af forbemeldte Eggert Ólafsen,* 2 volumes. [An appendix on Icelandic plants (*Flora Islandica*) by Johan Zoëga is at the end of Volume 2)]. Sorøe.

Paulli, S. (1648). *Flora Danica,* Kiøbenhafn.

Svabo, J. C. (1959). *Indberetninger fra ein Reise udi Færøe* 1781-82. København.

Hæwenhnydele: an Anglo-Saxon Medicinal Plant

CAROLE P. BIGGAM

46 Hamilton Avenue, Glasgow G41 4HD

Summary

The Old English plant-name, *hæwenhnydele*, occurs in herbal and medical texts and in glossaries containing translated Latin plant-names. Where it is linked with a Latin name, that name is always *Herba Britannica*, a cure for scurvy. Some scholars, rather naively assuming that the two names must refer to the same plant, have thought the identity of *hæwenhnydele* almost obvious, whereas others, knowing the frequently garbled accounts of herbal cures inherited by the Anglo-Saxons, have despaired of ever identifying it. An Anglo-Saxon translator, working on the Old English version of the Latin Pseudo-Apuleius, inherited an account of *Herba Britannica* which was a confusion of two different plants, compiled from several sources. The information available to him is discussed, in an effort to understand how he made an identification. Evidence such as the synonyms attached to the plant entry, the etymology of *hæwenhnydele*, and the illustration of the plant is presented. There is also a brief discussion of the two recorded occurrences of the rural plant-name, *hawdod*, from the 16th and 18th centuries, and whether this name could be cognate with *hæwenhnydele*. In the Durham Glossary, another name, *vihtmeresvyrt*, is linked with *Herba Britannica* and *hæwenhnydele*. This name is discussed, and found to denote a well-known source of Vitamin C, the cure for scurvy. This paper demonstrates that the attempts of the Anglo-Saxons to identify *Herba Britannica*, and to link their guesses with real herbal cures, are an impressive rationalisation of an almost nonsensical Latin plant record.

The Anglo-Saxon plant-name, *hæwenhnydele*, occurs eight times in surviving records, with various spellings. It is found in two Latin-to-Old English glossaries: the Brussels Glossary (Wright, 1884, p. 296), and the Durham Glossary (Von Lindheim, 1941, p. 10). It also occurs twice in the medical compilation known as *Lacnunga (*Grattan & Singer, 1952, pp. 98,122) and it occurs four times in the Old English *Herbarium* (Cockayne, 1864-6, **I**, p. 374; De Vriend, 1984, pp. 7, 74). Wherever *hæwenhnydele* translates a Latin plant-name, that name is *Herba Britannica*.

The opinions of scholars as to the identity of this plant are far from unanimous. Cockayne interpreted it as English scurvy-grass (*Cochlearia anglica* L.) with a question mark (Cockayne, 1864-6, **I**, p. 127), but he also suggested hemp nettle, with a question mark, referring to the genus *Galeopsis* (Cockayne, 1864-6, **III**, p. 329). Von Lindheim (1941, p. 33) described *hæwenhnydele* as '...an expression difficult to explain', but Grattan & Singer (1952, pp. 87-88) described their identification of it with purple dead-nettle (*Lamium purpureum* L.) as 'fairly safe'. Bierbaumer wisely concluded that the identification was 'unclarified' (Bierbaumer, 1975-9, **II**, p. 59), but Hunt, from later mediaeval evidence only, suggested that the mystery plant was the cowslip (*Primula veris* L.), with another question mark (Hunt, 1989, p. 214).

I intend to approach this problem by trying to reconstruct the decision processes of an Anglo-Saxon translator faced with a need to identify *Britannica* in his Latin sources. By far the most information comes from the Old English *Herbarium*, a translation of the Latin *Herbarium Apulei Platonici* and associated works. This 4th century Latin herbal is a compilation from works by writers such as Dioscorides and Pliny the Elder, and also from herbal traditions which do not otherwise survive. The account of *Britannica* in the Latin *Herbarium* has been researched (Biggam, 1993, pp. 214-21), and found to combine Pliny's description of *Britannica* with accounts of betony (*Stachys officinalis* (L.) Trev.), as a result of the similarity of the Greek names for these two plants: *Brettaniki* and *Bettoniki*. It should always be remembered that the transmission of texts before the invention of printing was prone to error. A text from Greece, for example, may have been copied hundreds of times before an Anglo-Saxon saw it in a Latin translation. Each copying created errors, and perpetuated, altered, or corrected earlier ones, but, moreover, early copyists had no concept of intellectual property, and often inserted sections from other texts without acknowledgement.

At a date within the 8th to 10th century (Cameron, 1983, p. 149; De Vriend, 1984, p. xlii), an Anglo-Saxon scholar came to translate the *Herbarium* account of *Britannica* into Old English. He must have made his identification of the plant from some or all of the following points of information in his source text: the cures which it is said to effect, the instructions on how to collect, store, and use it in these cures, a list of its names in several languages, and an illustration.

The first point concerns the cures, and those which are listed in the Latin texts closest to the Old English version, are for mouth sores, painful and loose teeth, upset stomach, and *paralysis*. Clearly, if the Anglo-Saxon translator knew of a native plant which effected the same cures, that would influence his identification, but it would be unlikely since his source text had confused two different plants.

The earliest surviving source text for most of these cures occurs in Pliny's account of how, in the campaigns of Germanicus Caesar across the Rhine between AD 14 and 16, the Roman soldiers began to lose their teeth and control of their knee-joints (Pliny the Elder [ed. Rackham *et al.*], 1969, **VII**, p. 151). They were suffering from scurvy, a Vitamin C deficiency disease. The Frisians, who were allies of the Romans at this time, told them to eat *Britannica*, and they were quickly cured. From this account, it is evident that, whatever *Britannica* was, it contained quantities of Vitamin C. Recent research on Pliny's narrative supports earlier suggestions that one of the best candidates for this plant is English scurvy-grass (*Cochlearia anglica*) (Biggam, 1993, pp. 212-4), but we should not assume that the Anglo-Saxon translator of the *Herbarium* made the same identification.

Although Pliny's encyclopaedia was known in Anglo-Saxon England, the translator appears to have been unaware that *Britannica* occurred in it, as can be seen from his translation of the term, *paralysis*. The word was a borrowing from Greek *paralisis* which means, literally, 'a loosening at the side,' but the Latin term was less specific, and was used of various types of impaired muscle control. Unaware that this cure originated in Pliny's account of collapsing knees, the Anglo-Saxon translator appears to have analysed the Greek word, since he translated *paralysis* as *sidan sar*, 'affliction of the side', which is much closer to the Greek meaning than the Latin. Thus the evident scholarship of the translator weakened the link with scurvy.

The information from the source text concerning the collection, storage, and use of the plant would only have helped the translator to identify it if exactly the same procedures were followed in his own medical traditions. There is little evidence on this point, but it seems unlikely that the very specific instructions (De

Vriend, 1984, pp. 75-7), would coincide with traditional English procedures.

The translator also had the evidence of the synonyms, the names for *Britannica* in several languages. The lists of synonyms in these early herbals are of considerable antiquity, and often contain names in non-European languages, some of which were already extinct by the mediaeval period. Under these circumstances, considerable distortion resulted, so that the names, as they were intended, were probably meaningless to our translator. However, I suspect that an error in one of them was used by him as a clue to the plant's appearance.

The Latin text informed the translator that 'the prophets' called this plant *caeluros* (Howald & Sigerist, 1927, p. 71). This is an error for a name which appears in other Latin manuscripts of this text as *aeluros* or *eluros*, but a translator with only one manuscript to consult would not know this. I believe he connected the erroneous name with Latin *caeruleus* 'blue', and this introduces the first element of the Old English name *hæwenhnydele*.

Hæwen is one of several problem words in Old English colour semantics. The Old English colour system differs considerably from both the Middle English and the Modern English systems, and this necessitates care in translation. Dictionary definitions of *hæwen* usually involve 'blue, purple, azure, green', and 'discoloured', but recent research has shown that the usual meaning of *hæwen* is 'blue', followed by 'grey' (Biggam, 1993, pp. 189-339). Thus the name *caeluros*, understood as *caeruleus*, is compatible with the Old English word *hæwen*.

The Anglo-Saxon translator, therefore, had only two clues to help him identify *Britannica*: the colour blue, and the illustration (Fig. 1). He did not make wild guesses at the plant identifications. Where he did not know, he left a blank in the manuscript where the Old English name was required, and this was done in 41 of the total 185 plant entries in the *Herbarium* (Cameron, 1993, p. 63). This suggests that, in the case of *Britannica*, he believed he had identified the plant. Our final clue occurs in the second element of the name used, *-hnydele*.

Hnydele appears in the surviving manuscripts with and without *n*, but it has been argued elsewhere that the form with *n* is probably more accurate (Biggam, 1993, pp. 229-31). It is suggested here that the ultimate origin of *-hnydele* is the Indo-European root **nē̆-* 'to sew', which gave rise to two groups of words in Indo-European languages. The first group involved the concept of the sewing instrument, such as Old High German *nâdela* 'needle', and the second group involved the concept of the sewing material, such as Old Norse hnotha 'a ball of thread'. Both groups produced related words which stressed the shape of these objects, so that Modern Icelandic *hnúthur* 'a knob, a head of a pin', for example, stresses the spherical shape of a ball of thread. A Modern English example is the dialectal *noddle*, 'head'. This concern with heads suggests a connection with the globular structures evident in the manuscript illustration of *hæwenhnydele*. Since *-hnydele* probably has a diminutive ending, it is suggested that *hæwenhnydele* means 'little blue heads'.

There are, of course, several plants which would suit this name, but two of them are supported by further evidence. The first plant, devil's-bit scabious (*Succisa pratensis* Moench) is supported by the manuscript illustration. The resemblancc is so noticeable that I accepted this identification before further evidence emerged (Biggam, 1993, pp. 233-4), and it was gratifying to find that

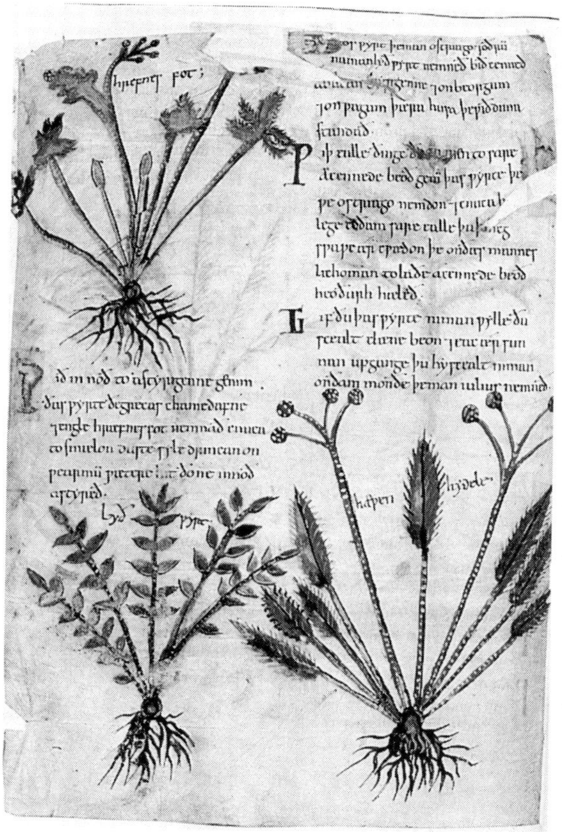

Fig. 1. *Hæwenhnydele* (below right) in MS British Library, Cotton Vitellius C. III, folio 26v. *By permission of the British Library.*

botanists at the Edinburgh symposium also saw the similarity. I am grateful to them for their valuable opinions. However, there is also evidence, consisting of two popular names, which supports an identification with the cornflower (*Centaurea cyanus* L.).

The first name appears in Hunt's list of popular names for the cornflower (Hunt, 1989, p. 145), occurring in the forms *nydel, nedul, nedull,* and *nedyll* in three 15th century herbal manuscripts. These words are etymologically acceptable as descendants of *-hnydele.*

The second popular name is *hawdods,* which occurs in only two recorded instances. It can be found in Fitzherbert's *Book of Husbandry,* published in 1534, in a list of the main weeds which grew in the crops of the day (Fitzherbert [ed. Skeat], 1882, p. 30). Fitzherbert describes the plant as having a blue flower. *Hawdods* occur again in the diary of John Hobson, who lived in Yorkshire in the early 18th century, and was told that this plant grew in the fields in summer (Morehouse, 1877, I, p. 296). Britten and Holland suggest *hawdods* are cornflowers (Britten & Holland, 1886). In the north of Britain, Old English *hæwen* became *haw,* which still occurs in Scots (Robinson, 1985). It is also possible to interpret *-dods* as 'heads', since the OED2 lists *dod* and *dodd* as having a primary sense of 'rounded head' (Simpson & Weiner, 1989). The etymology of *hawdods,* therefore, appears to be identical with that of *hæwenhnydele* with the exception of the diminutive.

It is clear that the devil's-bit scabious theory stresses the manuscript illustration, and the cornflower theory stresses the evidence of popular names, but it has to be admitted that both types of evidence are somewhat fragile. First of all, the illustrations may not have been taken seriously by the Anglo-Saxon translator. He must have been aware that the text sometimes contradicted features of the illustrations, and he probably suspected, quite rightly, that some of them had become attached to the wrong plants. He knew, furthermore, that artists squashed or stretched features of the plants according to the space available in the manuscript, since the text was generally written first, and he must have noticed that even recognizable illustrations were stylized and often fanciful.

If the illustration is of dubious value, what about popular names? It is well known that popular plant-names often denote different plants in different parts of the country, and can be used of more than one species in a single district. Grigson shows, for example, that devil's-bit scabious and cornflower are both known by the following names in different regions: bachelor's buttons, blue bobs, blue bonnets, blue buttons, and blue cap (Grigson, 1955).

It would appear, therefore, that the translator did one of two things after deciding that the plant had blue flowers. He was either prepared to accept the apparent depiction of devil's-bit scabious, or he noticed the distinctive little heads in an illustration which he otherwise distrusted, and was reminded of his name for the cornflower. It is assumed that he did not make up the name to fit the evidence, since he could have done that for all the plants he left without Old English names. Whichever identification is correct, it can be seen that, as a result of confusions in textual transmission, the very effective cure of English scurvy-grass for the disease of scurvy was lost to the Anglo-Saxons in this particular text.

Finally, it is interesting to look briefly at a separate Anglo-Saxon attempt to identify *Britannica,* which *did* result in an effective cure. In the Durham Glossary, *Britannica* is translated by *hæwenhnydele* and, also, by *vihtmeresvyrt* (Von Lindheim, 1941, p. 10). The manuscript is a 12th century Anglo-Norman copy of an earlier Anglo-Saxon manuscript, the copy having been made by a scribe who was not entirely familiar with Old English, and who consistently wrote *v* for *w*, and altered *æ* to *e* or *ea*. Thus, the correct Old English name should be *wihtmereswyrt* or *wihtmæreswyrt*. There is more than one possible interpretation of this name, but it has been argued elsewhere (Biggam, 1993, 236-7) that the best one is 'the plant which grows in clear pools'.

Certain evidence from glossaries suggests that the identification should be a cress. There are several entries in Latin-to-Old English glossaries which translate Latin *brittia* with Old English *cærse* (Wright, 1884, **I**, p. 271; Hessels, 1890, p. 25), and it seems likely that *brittia* could have been mistaken for an abbreviation of *Britannica*, while Old English *cærse* means 'cress.' It should be noted that early glossaries were often compiled by collecting translations of Latin words from several different manuscripts, so it looks as if the Durham Glossary has recorded the efforts of two different translators to interpret *Britannica*, one deciding on the cornflower or devil's-bit scabious, and the other on a cress.

The thought processes of the second translator cannot be reconstructed, since no full text with this translation survives, but it is interesting that Pliny's *Britannica* may be English scurvy-grass, and *grass* in this plant-name is a distortion of *cress.*

Since the second translator appears to have had more information about *Britannica* than the first, it is tempting to suggest that *wihtmereswyrt* originated as a gloss to *Britannica* in Pliny's account of the Romans in Frisia. Since the cresses are well-known for their Vitamin C content, this cure for scurvy was as good as any modern tablet.

References

Bierbaumer, P. (1975-9). *Der botanische Wortschatz des Altenglischen*, 3 vols. Bern: Lang.

Biggam, C.P. (1993). *A Lexical Semantic Study of Blue and Grey in Old English: a Pilot Study in Interdisciplinary Semantics*, Ph.D. Thesis, University of Strathclyde.

Britten, J. and Holland, R. (1886). *A Dictionary of English Plant-Names*, 618 pp. London: Trübner.

Cameron, M.L. (1983). The Sources of Medical Knowledge in Anglo-Saxon England. *Anglo-Saxon England* **11**, 135-155.

Cameron, M.L. (1993). *Anglo-Saxon Medicine*, 211 pp. Cambridge: Cambridge University Press.

Cockayne, O. (1864-6). *Leechdoms, Wortcunning, and Starcraft of Early England*, 3 vols. London: H.M.S.O.

De Vriend, H.J. (1984). *The Old English 'Herbarium' and 'Medicina de Quadrupedibus'*, 403 pp. London: Oxford University Press.

Fitzherbert, A. (1882). *The Book of Husbandry, Reprinted from the Edition of 1534* (ed. W.W. Skeat), 167 pp. London: Trübner.

Grattan, J.H.G. & Singer, C. (1952). *Anglo-Saxon Magic and Medicine Illustrated Specially from the Semi-Pagan Text 'Lacnunga'*, 234 pp. London: Oxford University Press.

Grigson, G. (1955). *The Englishman's Flora*, 478 pp. London: Dent..

Hessels, J.H. (1890). *An Eighth Century Latin-Anglo-Saxon Glossary Preserved in the Library of Corpus Christi College, Cambridge (MS No. 144)*, 226 pp. Cambridge: Cambridge University Press.

Howald, E. & Sigerist, H.E. (1927). *Antonii Musae 'De herba vettonica liber', Pseudoapulei 'Herbarius', Anonymi 'De taxone liber', Sexti Placiti 'Liber medicinae ex animalibus...'*, 297 pp. Leipzig: Teubner.

Hunt, T. (1989). *Plant Names of Mediaeval England*, 334 pp. Cambridge: Brewer.

Morehouse, H.J. (1877). *Yorkshire Diaries and Autobiographies in the Seventeenth and Eighteenth Centuries*, 2 vols. Durham: Andrews.

Pliny the Elder (1969). *Natural History* (ed. H. Rackham, W.H.S. Jones, & D.E. Eichholz), 10 vols. London: Heinemann.

Robinson, M. (1985). *The Concise Scots Dictionary*, 820 pp. Aberdeen: Aberdeen University Press.

Simpson, J.A. & Weiner, E.S.C. (1989). *The Oxford English Dictionary*, 2nd edition, 20 vols. Oxford: Oxford University Press.

Von Lindheim, B. (1941). *Das Durhamer Pflanzenglossar, Lateinisch und Altenglisch*, 81pp. Bochum-Langendreer: Pöppinghaus.

Wright, T. (1884). *Anglo-Saxon and Old English Vocabularies*, 2nd edition (ed. R.P. Wülcker), 2 vols. London: Trübner.

The Physicians of Myddfai: The Welsh Herbal Tradition

HEATHER M. HENDERSON, M.R.H.

80 Queens Road, Penarth, South Glamorgan, CF64 1DJ

Summary

The knowledge of medicine in Wales goes back to ancient Celtic times when it was handed down orally through many generations. During the 13th century, a Prince of South Wales, Rhys Gryg, instructed his court physician, Rhiwallon, to gather together the knowledge of medicinal plants and their methods of application and to write it down in Welsh. This would enable others to make use of the information both then and during the centuries that followed. Rhiwallon lived at Myddfai, a tiny village in Mid-Wales. He had three sons, who worked as physicians to the court and the neighbouring area. They became famous because of their skills in medicine and in due course a legend grew up around them. The countryside around Myddfai is still very rich in medicinal plants. They include *Valeriana officinalis, Filipendula ulmaria, Scrophularia nodosa, Digitalis purpurea, Plantago* spp., and *Achillea millefolium*; the uses to which the Myddfai Physicians put these plants, and others, are described. 'Magical' plants such as *Verbena officinalis* and *Viscum album*, which have Druidic connections, are also discussed; some of these are still used today by herbalists. Many of the prescriptions of the Physicians are thus as relevant today as they were in the far past.

Introduction

Myddfai is a very small village in Mid-Wales, at the foothills of the Black Mountain or Carmarthen Vans. The nearest town is Llandovery; from there, the River Tywi flows down the valley to join the sea beyond Carmarthen town. According to legend, the Mother of the Physicians of Myddfai was the fairy woman Nelferch, who came from the 'Other World' and used to rise out of a nearby lake, Llyn-y-Fan-Fach; the first of the Physicians was reputedly her son by an earthly father. She instructed them in the arts of healing (Grigson, 1975: p. 19).

In reality, the Physicians would have been learned men, some of them attached to the courts of local Princes and Lords. They were granted land to enable them to 'practice their art' for the benefit of the local community. Their prescriptions began to be written down, in Welsh, in the early 13th century by Rhiwallon, court physician to Rhys Gryg, a Prince of South Wales. The writings represented the knowledge of medicine which had passed down orally for hundreds of years before Rhiwallon. The Doctors certainly had a wide, detailed knowledge of local plants and how to use them. The value of these written prescriptions was to the future generations of Doctors who were able to use them as a base from which to work, and to add to, which was certainly done up to the 18th century.

The local Doctors not only had a wide knowledge of the native plants but also of those from abroad such as myrrh (*Commiphora abyssinica* (Berg.) Engler) and frankincense (*Boswellia sacra* Flueckiger) which were both used frequently in ointments, as well as canella bark (*Canella winteriana* Gaertner), a tonic and stimulant from the West Indies. These were obviously imported but must have been readily available to be included in so many of the prescriptions, although some, such as *Canella*, would not have been available to the earliest of the Physicians, whose prescriptions began to be written down in Pre-Columbian times.

Some of the prescriptions used compounds such as copper sulphate in healing ointments; quicksilver (mercury) was used in rubefacients for rheumatism. These substances are no longer recommended! This is what makes the writings so interesting as they are a living record from the oral tradition to at least the mid-18th century, when William Bona wrote: 'I William Bona have transcribed this work from the book of John Jones, the Physician of Myddfai, who was the last of the descendants of the Physicians of Myddfai, Anno Christii 1743'; 'and I, Iolo Morganwg have rewritten the same carefully from the book of the above William Bona now in the possession of Thomas Bona, Physician, of the parish of Llanfihangal Iorwerth, in the County of Carmarthen, in the year 1801; and with old Howel the Physician I say the grace of God attend it.'

The prescriptions

Obviously some of the prescriptions are of dubious value today, especially those using mineral extracts and various parts of animals. However the largest number are purely herbal and many are still in use. It is interesting that some of the obvious medicinal herbs are not mentioned, or only rarely so. For example, meadowsweet (*Filipendula ulmaria* (L.) Maxim.) is only mentioned three times, yet it is a source of salicylic acid, natural aspirin, and so good for pain relief in rheumatism and for lowering temperature. (Another source of aspirin is willow, *Salix* L. spp.; *cf.* Bennett, 1994, this volume - *Editor.*). It is also used for digestive problems and is known as the 'herbal bicarbonate of soda'. Meadowsweet grows in profusion in Mid-Wales due to the damp climate and the prevalence of natural farming methods. Another virtually overlooked species is dandelion (*Taraxacum officinale* Wiggers, agg.), which is only mentioned five times. It is an excellent diuretic and digestive tonic. The knowledge of these 'simples' would have been held by the local wise women, to whom people would have gone first with their common complaints, and certainly the poor peasants would have used this knowledge. It is probable that Rhiwallon and his successors thought these plants too ordinary and their uses too well known to be written down in such a book. Gerard (1597), in his 16th century herbal, certainly made references to such 'common plants' that were too well known to be included.

Wounds were very common and there are many prescriptions for ointments to treat anything from an infected, running sore to an open wound. Frankincense and myrrh resins were often included in these ointments, as they are today, along with various waxes and fats to set the ointments.

Nettle (*Urtica dioica* L.) roots, goosegrease and honey were used as a drawing plaster to remove foreign objects. Feverfew (*Tanacetum parthenium* (L.) Schultz-Bip.), plantains (*Plantago major* L., *P. lanceolata* L.) and sage (*Salvia officinalis* L.) were made into an ointment for bruises. For boils and bleeding sores, frankincense (*Boswellia sacra* Flueckiger) and mastic (*Pistacia lentiscus* L.) were boiled with lard and beeswax and flowers of sulphur added. A general healing salve comprised plantains (*Plantago major* L., *P. lanceolata* L.), scarlet pimpernel (*Anagallis arvensis* L.) and tansy (*Tanacetum vulgare* L.) boiled in lard with frankincense added. Yarrow (*Achillea millefolium* L.) was another important wound healing herb and was used as such by the ancient Greeks for their battle wounds; this knowledge passed into Anglo-Saxon tradition *via* an Anglo-Saxon translation of the *herbarium* of Apuleius Platonicus (Grigson, 1975). Yarrow is both astringent and antiseptic and has wide use today for a number of circulatory problems and infections.

Digestive problems were as common then as they are now, although obviously the diet was not as varied as ours, especially in the winter. Wild carrot seed *(Daucus carota* L.) was recommended for flatulence and is still used. Other herbs used to treat digestive problems were fennel (*Foeniculum vulgare* Miller), dill (*Anethum graveolens* L.), the mints (*Mentha* L. spp.), mugwort (*Artemisia vulgaris* L.), parsley (*Petroselinum crispum* (Miller) Nyman ex A .W. Hill), tansy (*Tanacetum vulgare* L.) and juniper berries (*Juniperus communis* L.).

The root of *Berberis vulgaris* L., or the bark of the stem, was boiled in ale to treat jaundice. (This stems from Coles (1657) who wrote that the yellow inner bark of the roots and branches bore the signature of jaundice. - *Editor*). *Berberis vulgaris* is still used for inflammation of the gall bladder, for improving liver function and as a bitter tonic. Centaury (*Centaurium erythraea* Rafn) was another widely used tonic, particularly where there was also heart disease. It is again a bitter tonic and was used all over Europe to treat anaemia and liver problems. It also calms the nervous system and helps depression, so indeed it would have a beneficial effect on the whole body.

Butterwort (*Pinguicula vulgaris* L.) was made into a complicated concoction for treating 'the stone'. *i.e.* the kidneys, was called 'the blue confection of the stone'. Butterwort is fairly common on the acid moorland around Myddfai.

One of the most popular herbs in the prescriptions for treating respiratory problems was sage (*Salvia officinalis* L.), being mentioned 51 times. Sage is very antiseptic, especially for throat conditions. Elder flowers (*Sambucus nigra* L.) were used for their anti-catarrhal properties and are good for relieving the symptoms of the common cold and hay fever. Ground-ivy (*Glechoma hederacea* L.) was used as an expectorant and astringent for bronchitis, and hyssop (*Hyssopus officinalis* L.) for a tight chest and asthma as it has antispasmodic properties. Rosemary (*Rosmarinus officinalis* L.) was frequently suggested for clearing phlegm from the lungs, both as a decoction and by inhaling the vapour which would contain the essential oils. Elecampane (*Inula helenium* L.) must have been grown in the Physicians' gardens, as although not native it is mentioned in 13 prescriptions for lung conditions. It was also used externally, as its old name of scabwort testifies. *Inula helenium* is active against the tuberculosis bacillus and today is often used to treat chest infections. Figwort (*Scrophularia nodosa* L.) was used for eczema and psoriasis, to reduce the irritation. Another important skin herb was greater burdock (*Arctium lappa* L.), especially for psoriasis for which it is still used.

Valerian (*Valeriana officinalis* L.) is a common plant in Mid Wales. Its root is a natural tranquilliser, but has a very unpleasant taste. It was used, along with stinging nettle (*Urtica dioica* L.) and dwarf elder (*Sambucus ebulus* L.) to relieve pain and paralysis. Dwarf elder, or danewort, has never been frequently used (it is, at least now, an uncommon species) but the leaves are both diuretic and purgative.

The Physicians used a wonderful anaesthetic mixture before carrying out operations. This was composed of orpine (*Sedum telephium* L.), which was found in the hedgerows around Myddfai, eryngo (probably *Eryngium maritimum* L.), opium poppy (*Papaver somniferum* L.), mandrake (*Mandragora officinalis* L.), ground-ivy (*Glechoma hederacea* L.), hemlock (*Conium maculatum* L.) and great lettuce (*Lactuca virosa* L.). This was mixed with clean earth and dropped into the patient's nostrils. To wake him up afterwards a sponge dipped in vinegar was held to the nostrils.

Sacred Herbs of the Druids and their use by the Physicians

The Druids, whose knowledge of medicinal plants is the foundation of the writings of the Physicians of Myddfai, had a number of sacred plants that were used extensively in their medicine (Rutherford, 1993). Some are still being used today. A good example is mistletoe (*Viscum album* L.). It was recommended by the Physicians.' In any case of bodily debility, whether in the nerves, joints, back, head or brain, stomach, heart, lungs or kidneys, take three spoonfuls of the decoction and mix with boiling water, ale, mead or milk; then add to a good draught thereof a spoonful of the powder, which should be drunk in the morning fasting.' Considerable details are given about the collection of both the berries and the green twigs. The preferred plants grew on oak (*Quercus robur* L., *Q. petraea* (Mattuschka) Liebl.). The best places to collect them from were Monmouthshire, Somerset, Brecknockshire and Shropshire. Mistletoe is used today to lower blood pressure and for helping headaches associated with hypertension. It is very effective at relaxing the nervous system and is also being investigated for anti-cancer properties.

Oak itself, and vervain (*Verbena officinalis* L.), were also very important plants to the Druids [for vervain, cf. Grigson (1975) p. 339; Rutherford, 1993: p. 117-118 - *Editor*]. Today we use oak for its astringent and tonic properties for the digestive system. Vervain is also a tonic herb both for the digestion but more importantly for the nervous system. The Physicians used it as a general 'cure-all', along with another sacred herb, betony (*Stachys officinalis* (L.) Trev. St. Leon). The latter had many virtues attached to it; it contains three alkaloids that are also found in yarrow (*Achillea millefolium*), therefore it has similar wound healing properties. Internally, betony stimulates the circulation and will relieve headache and migraines.

Conclusion

Whilst some of the ingredients in the prescriptions seem strange and some even repugnant, not to mention some of the methods of application, the writings of the Physicians are very valuable, not only from the historical aspect but also from the point of view of medicine today, where there is research being carried out on plants known to have curative properties, for the drugs of the future.

Warning

Many of the plants mentioned in this paper are poisonous. Information given in the paper is not intended to be used instead of, or in replacement for, any medication prescribed. In case of illness, contact your local Doctor or a fully qualified Medical Herbalist.

References

Bryce, D. (ed.) & Pughe, J. (transl.). (1987). *The Herbal Remedies of the Physicians of Myddfai*, 151 pp. Felinfach, Lampeter, Dyfed: Llanerch Enterprises.

Coles, W. (1657). *Adam in Eden: or, Nature's Paradise: the History of Plants, Fruits, Herbs and Flowers,* [12], 629, [22] pp. London: J. Streater, for Nathaniel Brooke.

Gerard, J. (1597). *The Herbal or Generall Historie of Plantes*, [20], 1392, [67] pp. London: John Norton.

Grigson, G. (1975). *The Englishman's Flora*, 542 pp. Frogmore, St Albans: Granada Publishing (Paladin).

Rutherford, W. (1993). *Celtic Lore. The History of the Druids and Their Timeless Traditions*. 224 pp. London: HarperCollins Publishers (The Aquarian Press).

Castles and Abbeys in Wales: Refugia for 'Mediaeval' Medicinal Plants

ANN CONOLLY

25 Brocks Hill Drive, Oadby, Leicester LE2 5RE

Summary

The restriction to the walls and immediate vicinity of Welsh castles, abbeys and other ancient settlement of a group of species known to have medicinal properties, coupled with the absence of these species from apparently suitable intervening terrain, suggests relict status from former cultivation. On site archaeological evidence of relevant seed or other macro-remains is sparse from these sites. Arguments in favour of such lengthy survival at these 'ancient' locations are discussed, as is the alternative view that only at such sites are suitable habitat requirements met. Species considered include *Sambucus ebulus*, *Hyoscyamus niger*, *Verbena officinalis* and *Lamium album*.

Introduction

In Wales and the Marches a characteristic group of synanthropic species grows on or close to castles and abbeys or other ancient settlement. The restriction to such sites is often marked and for the rarer species their isolation great. Many of these species have medicinal or other herbal properties suggesting a purposeful importation (and probable cultivation) and maybe a long survival '*in situ*' as relicts. My first awareness of this association came from references (Hyde & Wade, 1957) to the rare *Diplotaxis tenuifolia* (L.)DC. growing on Harlech Castle (one of only two sites in Merioneth), *Descurainia sophia* (L.)Webb ex Prantl on Denbigh Castle and *Marrubium vulgare* L. at Gogarth Abbey (Great Orme). Then, exploring Wales and the Marches for *Lamium album* L., it became apparent that at some sites there was an association with 'ancient monuments' (Flint Castle, Whitland Abbey, Caerleon Roman amphitheatre), and at others with un-modernized farms and remote country villages, often there accompanied by *Chelidonium majus* L., *Ballota nigra* L. or *Malva sylvestris* L. (Conolly 1972). Were these, too, relicts of a former 'cultivation'? , or were there particular attributes of soil, climate or some local natural factor to explain the linkage?.

The subsequent survey of some 50 castles (Table 1) and nearly 20 Abbeys or other 'monastic' sites (Table 2) throughout Wales and the Marches, ranged from Pembroke Castle and St Davids and Chepstow Castle and Tintern in the south, to Beaumaris and Flint Castles and Basingwerk Abbey in the north, and demonstrated the recurrence again and again of these special species, each site with its own admixture. Two major habitats are involved: firstly the walls and ruins and secondly, associated places other than walls, on ground nearby or in the vicinity of other ancient habitation. The relevant species may be either rare and of limited occurrence in Wales, or more generally characteristic and found more widely.

Table 1. Castles in Wales and the Marches

Abergavenny	Cilgerran	Harlech	Newcastle Emlyn
Aberystwyth	Clun	Haverfordwest	Ogmore
Beaumaris	Coity	Kidwelly	Pennard
Brecon	Conway	Laugharne	Penhow
Bridgend	Criccieth	Llandovery	Pembroke
Builth Wells	Denbigh	Llanstephan	Raglan
Caerphilly	Dinas Powys	Llawhaden	Rhuddlan
Caernarfon	Dinefwr	Lougher	Skenfrith
Cardiff	Dolbadarn	Ludlow	Tenby
Carmarthen	Dolwyddelan	Manorbier	Tretower
Carreg Cennen	Dryslwyn	Monmouth	Weobley
Chepstow	Flint	Montgomery	White Castle

Table 2. Abbeys and other monastic sites in Wales and the Marches

Abbey Dore	Kenfig	Puffin Island* (Penmon Priory)	St Tudwal's*
Basingwerk	Llandewi (Gower)	Strata Florida	Talley Abbey
Bardsey*	Llanthony	St Dogmaels	Valle Crucis
Caldey*	Margam	St Davids	Whitland
Cymmer Abbey	Neath Abbey		

*Monastic Islands

Species of castle and abbey walls

Species of limited occurence, often rare throughout Wales (Table 3a)

Diplotaxis tenuifolia (Harlech, Carew Castle (Davis, 1970), Haverfordwest Tenby town walls), *Descurainea sophia* (Denbigh Castle 1898, St David's 1869) and *Sisymbrium orientale* L. can also turn up in waste places as recent (alien) casuals often near the coast as befits their continental or southern European extraction. But the *Sisymbrium* persists, doubtless long-established, in two coastal towns in Lleyn (Fig 1a). Both the alien *Senecio squalidus* L. and *Buddleja davidii* Franchet are recent non-herbal invaders, though 1898 (Cardiff Castle) is an early date for the former. The undoubted native *Cochlearia danica* L. finds an unusual inland habitat on the castle walls of Newcastle Emlyn and Kidwelly. *Verbena officinalis* L. (perched on Skenfrith Castle wall) and *Salvia verbenaca* L. on the ruins of Rhuddlan Castle (Wynne, 1993) are well-known medicinal herbs. *Brassica oleracea* L. on Chepstow Castle, and at Tenby both on the Castle and on the cliffs nearby, raises the question whether wild plants have 'come in' from the local cliffs, or are escapes from cultivation reverted to 'wild-like' habit, as one supposes must be the 'Wild Cabbages' on the cliffs of San Francisco Bay?. The *Diplotaxis* and *Sisymbrium* seem not to be 'medicinal', but 'introductions' nevertheless. *D. sophia* has declined since its 17th century frequency (Salisbury, 1961).

Table 3. Species found on castle/abbey walls

a) ***of limited occurrence***

Diplotaxis tenuifolia (Harlech, Carew, Haverfordwest)
Descurainea sophia (Denbigh 1898, St Davids 1869)
Sisymbrium orientale (Raglan, Haverfordwest)
Brassica oleracea (Tenby 1775 on, Chepstow)
Antirrhinum majus (Cardiff, Monmouth, Valle Crucis)
Salvia verbenaca (Rhuddlan, Tenby, Penmon Priory)
Verbena officinalis (Skenfrith)
Reseda luteola (Montgomery, Raglan, Neath)
native:
Cochlearia danica (Newcastle Emlyn, Kidwelly)
recent introductions:
Senecio squalidus (Abergavenny, Neath, Cardiff 1898)
Buddleja davidii (Bridgend, Neath)

b) ***of wider occurrence***

Centranthus ruber
Erysimum cheiri
Cymbalaria muralis
Parietaria judaica
native:
Arabidopsis thaliana
Erophila verna
Polypodium cambricum
Geranium lucidum
Saxifraga tridactylites

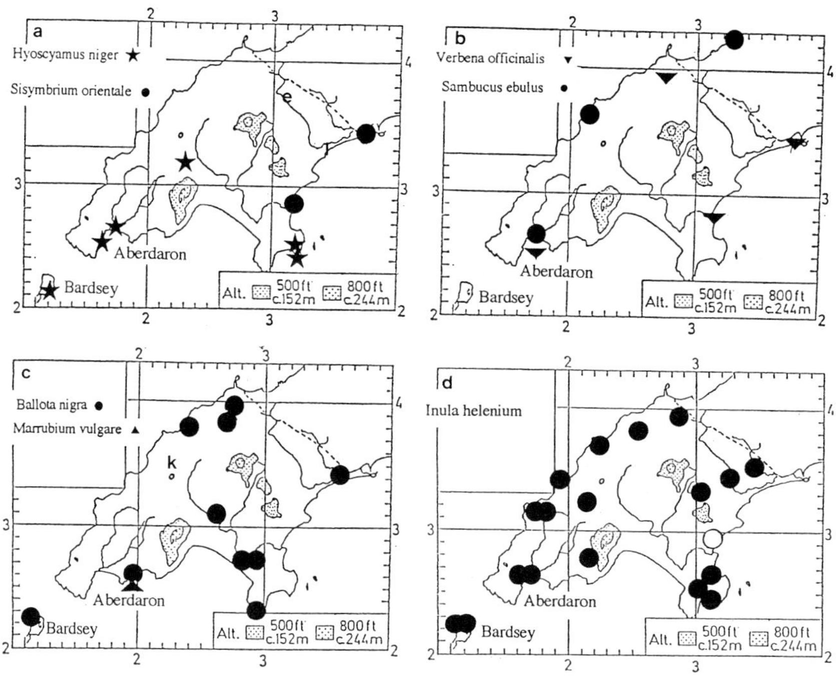

Fig.1. Medicinal plants in Lleyn

Characteristic species but of wider occurrence (Table 3b)

The non-herbal native species listed as characteristic on many castle walls are regarded as opportunists; *Erophila* sp. (*verna* s.l.) and *Saxifraga tridactylites* L. perhaps finding crumbling mortar, as inland on Raglan Castle, equivalent to calcareous coastal sand. *Polypodium cambricum* L. is notable on Criccieth and Dinefwr Castles (Chater, pers. comm. 1978). *Erysimum cheiri* (L.) Crantz (*Cheiranthus cheiri* L.) is typical too of ancient walls other than those of castles (where it always has golden flowers) and *Centranthus ruber* (L.) DC., with a wider range of substrate, are of long standing. Both are of Mediterranean origin and in Wales are more frequent coastally and on limestone. *Parietaria judaica* L. is perhaps the most constant of the Castle/Abbey wall species with *c*.75% frequency and it can be abundant - as at one time over Bardsey's ruined Abbey tower. Though widely found on old cottage and village walls throughout Wales its rarity on natural rocky outcrops and sea cliffs makes one wonder whether these might be secondarily derived by spread from local ruins? This would imply an ancient inbringing from southern or south-western Europe for this species with known pharmacological properties. Or has a rare coastal native dramatically 'exploded' its range inland when the human population took to building walls?

Species associated with castles, abbeys or other ancient habitation growing on the ground (not walls).

Species of restricted occurrence. (Table 4a)

Species in this group grow either in the immediate vicinity of a castle or abbey, or, if otherwise, then always close to human habitation. They are all known medicinal herbs. Some (*Atropa belladonna* L. *Marrubium, Verbena*) are considered native especially in southern, calcareous parts of England, but in Wales they are rare and closely 'tied' to human settlement: doubtless as relicts of a past (or continuing) keeping of herbal remedies to hand. This is demonstrated too by the frequency of Tansy (*Tanacetum vulgare* L.) on farmyard wall-tops, or Elecampane (*Inula helenium* L.) also Manorbier Castle (Salisbury, 1961) and Comfrey (*Symphytum officinale* L.) in hedge or orchard in, for example, rural Lleyn (Fig. 1d). *Atropa belladonna* has a mere score or so of sites in Wales including near Tintern Abbey, Hawarden Castle and Whitefriars, Cardiff (Trow, 1911). *Marrubium vulgare*, though with quite a few localities on the coastal limestone of north Wales (considered by Salisbury (1961) to have spread there from Gogarth Abbey) and south Wales, has only a scatter elsewhere: including Weobley Castle and (the monastic) Caldey Island in the south (Hepper, 1954; Rees 1950), and Penmon Priory (Anglesey) in the north (Roberts, 1982). In west Lleyn it now grows at a single farm - a 13th century holding of the Abbey of Enlli (Bardsey) (Johns, 1960; Jones-Pierce, 1963) accompanied by *Ballota nigra, Conium maculatum* and *Malva sylvestris* (Fig. 1c). *Verbena officinalis* is more frequent in Pembrokeshire and on limestone: it is known at Skenfrith Castle, Kidwelly Castle (May, 1967), Roman Pennal (Merioneth) (Benoit & Richards, 1963), and Beaumaris (Roberts, 1982). In west Lleyn it is long-persistent in merely four villages (Fig. 1b). *Sambucus ebulus* L. has a paucity of sites in Wales, all remote isolated stands; that beside Llandovery Castle (known since 1773) must be 30km from the next nearest stand. Yet it sets seed here and birds (Lesser Whitethroat, Ridley, 1930) do eat the berries. In Lleyn there are just four colonies strung along the pilgrim route to Bardsey Island. (Fig. 1b). So what prevents it spreading? Do all sites represent a planting of imported 'roots'? Salisbury (1961) suggest a Saxon introduction. *Cynoglossum officinale* L. (in sand Kenfig Castle) is regarded by Svensson &Wigren (1986; 1990) as cultivated in the Middle Ages because of its isolated occurrences on convent ruins. *Hyoscyamus niger* L. has a long history of persistence - often close to ruins, and often intermittently following disturbance of long-buried seed. It can be no coincidence that this potent drug plant grows on small Welsh off-shore islands favoured by early 'Celtic' hermits of the 6th century: Bardsey, Puffin Island, Caldey, as well as Steepholm. What better remedy to have on hand for storm-bound monks with toothache? In Lleyn (Fig. 1a), I suspect the *Hyoscyamus* on the Aberdaron beach (1961, 1966) to be water-borne from Bardsey, or from the nearby landing bay where sheep from the island are off-loaded; but Ridley (1930) doubts the effectiveness of the seeds 'one-day buoyancy'. The provenance of the inland root-crop weed could be seed contaminant, but the coincidence is curious that this field was once part of the grange-land (*mynachdy*) estate under the lordship of the Abbot of Enlli (Bardsey) (Jones-Pierce, 1963; Williams, 1990). Not only *Hyoscyamus* grows on these monastic islands, but a suite of medicinal herbs: on Bardsey *Artemisia absinthium, Inula helenium* and *Datura stramonium* L.) also Caldey(Rees, 1950); on Puffin Island *Smyrnium* and *Verbascum thapsus* (Lacey, 1957). And on the associated 'monastic' land in and near Aberdaron still grow *Marrubium, Verbena, Sambucus ebulus, Smyrnium* and *Datura*; and by Penmon Priory *Marrubium, Foeniculum, Reseda luteola* and at one time *Salvia verbenaca* (Roberts, 1982). Moreover on both Bardsey and Puffin Island the more widespread species, *Conium, Ballota, Malva sylvestris* and *Parietaria,* all potential medicinal species, also occur. A natural dispersal of such an assemblage of species with restricted mainland sites to these several islands would seem improbable. But whether first introduced by mediaeval monastics or subsequently, even recent times, cannot now be said without dated fossil evidence.

Table. 4. Species associated with castles and abbeys, growing in vicinity but not on the walls.

a) of restricted occurrence	*b) of wider occurrence*
Atropa belladonna (Tintern, Hawarden 1772, Cardiff)	*Ballota nigra*
Marrubium vulgare (Weobley, Gogarth Abbey Penmon, Caldey)	*Malva sylvestris*
Verbena officinalis (Skenfrith)	*Conium maculatum*
Sambucus ebulus (Llandovery 1773 on)	*Smyrnium olusatrum*
Hyoscyamus niger (Puffin Is., Bardsey Is., Caldey Is. Rhuddlan)	*Artemisia absinthium*
Myrrhis odorata (Valle Crucis, Bardsey Is.)	*Chelidonium majus*
	Lamium album
	Verbascum thapus

There is considerable macro fossil evidence for many of these medicinal plants (Table 4a) from England, but little from Wales: and mostly of Roman or Mediaeval age. *Atropa belladonna* recovered from Roman Caerwent and Carmarthen (town) has wide Roman and Mediaeval records from urban sites in England (Hall,

1986) and a mediaeval horde is reported from Elgin (Scotland) (Dickson, 1992). *Verbena* from Roman Caerwent, *cf. Sambucus ebulus* Roman Carmarthen, and *Foeniculum vulgare* also from Caerwent (Roman) and Tenby (Tudor) completes the Welsh list.

For *Hyoscyamus niger* there is an extensive list of Roman and Mediaeval finds from England, but from Wales a single 'pre-Roman-Iron Age' site near St Clears (Carms.), and two Roman ones - Carmarthen Town (3-4th century AD) and Prestatyn (Flints) (*c.* 1st-2nd century AD) where there is a modern (?pre-1930) record (Wynne, 1993), apparently the only coincidence between macrofossil and modern occurrence. Although actual survival at a specific site is speculative nevertheless longevity of buried seed is well authenticated as illustrated by recent reports in *B.S.B.I.News* (1988-1990) of *Atropa belladonna, Verbascum thapsus, Datura stramonium* and *Hyoscyamus niger* appearing after soil disturbance - but a local source improbable. *Hyoscyamus* is notorious for its intermittent appearance as on the monastic St Patrick's Isle, Isle of Man (Garrad, 1989). The most striking demonstration comes from Denmark where Ødum (1965, 1978) has not only observed (or reported) many of these medicinal and ruderal species growing on spoil heaps of current archaeological excavations (but not known in the vicinity), but even the spectacular germination, *in situ,* along freshly exposed profiles, of *Hyoscyamus niger* or *Verbascum thapsus.* Soil from a Cistercian monastery excavation grew a typical assemblage of *Hyoscyamus, Marrubium, Ballota, Malva sylvestris, Conium* and *Chelidonium.* Moreover, from extracted, dated soil he germinated seeds of many such species, giving evidence of viable 'survival' of, for example, *Hyoscyamus* seeds since AD 1300.

Species of wider occurrence (Table 4b)

The species in this group, like the previous, are typical of the immediate vicinity of castle or abbey, often on ground below the walls, and occasionally even on them (*Ballota nigra, Malva sylvestris, Verbascum thapsus*); they are also found over much of lowland Wales, but restricted, especially in the west, to the proximity of ancient dwellings, village or farm. Far commoner than the preceding group, and generally considered to be natives in Britain, they all have medicinal usage however, so that a former cultivation and modern restriction as village relicts, as Synnott (1979) supposes for Ireland, is surely also feasible for at least western Wales. *Chelidonium majus* L. invariably marks the site of old habitation, usually in company with others in this group. *Conium maculatum* and *Malva sylvestris* are the two next most frequently recorded species after *Parietaria* at the castle and abbey sites (Tables 1,2). The *Ballota* and *Malva* do best coastally and on limestone; and so does *Smyrnium*, a species from southern Europe, which spreads further from village centres coastally. It would appear that these species near the limit of their climatic range are more dependent on nitrogenous and phosphate rich soils, as Saarisalo-Taubert (1963) found in the southern Finnish town of Porvoo. And limestone outcrops, like mortared walls, give better drainage for species originally native in a drier climate.

Even *Lamium album* L. can be included in this group; similarly rare in the far west (and north) of Britain, in lowland western Wales it is never far from habitation and is characteristic of castle and abbey sites. It too is used as a herbal remedy even today; several instances of deliberate transfer of 'roots' for this purpose over a distance of *c.* 30 km have been confirmed, with a survival 'on site' of over 80 years. Does the curious coincidence with Roman stations in central Wales imply a Roman importation? or merely the availability of suitable phosphate-nitrate rich soil? That *Lamium album* and the rest of this group may have been originally brought into Wales intentionally seems possible - perhaps even by Romans.

There are more macrofossil records in Wales for this group than for the more exacting medicinal species already cited. The Roman site at Caerwent yielded an assemblage of *Ballota, Chelidonium, Conium,* cf *Malva* sp., *Smyrnium* and *Urtica urens* L. A few of these species have been recovered from other Roman sites in Wales: Carmarthen-*Chelidonium, Malva sylvestris* and *Urtica urens*; Whitton-*Conium* and cf *Lamium album*; Prestatyn - *Conium* and *Urtica urens.* There is only a single earlier dated find from Wales, an 'Iron Age' *Conium maculatum* from Collfryn (Montgomeryshire). But there are earlier dates for a good number of these species (with Welsh fossil records) from England - mostly from eastern sites. This applies similarly to the previous group, although *Marrubium vulgare* is an exception with no macrofossil records from Wales nor pre-Roman from any British archaeological deposit - all are Roman or Mediaeval. On the continent it is known first from Mediaeval times (Willerding, 1986). Does not this suggest a very late introduction?

The question remains whether the medicinal species listed have been intentionally introduced (at least in the first instance) or whether the restriction to isolated sites found today is the outcome of limited availability of appropriate habitat - of soil, drainage and other features - that the castles and other ruins and their surrounds provide for species from more southern parts of Europe and the Mediterranean. The repetition at so many castles, abbeys and similar sites of these species, the long distances between these sites with an absence from the intervening area, suggests a lack of ability to spread and disperse naturally. This is a feature in keeping with 'alien' species often transferred as 'roots' or cuttings. But modern practices have much reduced the occurrences (and increased the isolation) of many of the village 'herbals'. Wild plants will have been gathered locally and brought into the settlement area too, and a few of the imported species may have spread onto natural habitats providing equivalent 'advantages', such as limestone and coastal conditions, to those of the settlement sites.

It is hard to believe that an assemblage of these known medicinal species could have come together on a small monastic island or at a particular castle or abbey other than by a purposeful introduction by man. And such assemblages are repeated for some 70 Welsh sites, though each with a different set of species. Whether these species arrived contemporaneously with the occupation date as argued by Synnott (1979) for Ireland, is difficult to establish. In any case supplementary 'arrivals' (of different stock) or new species will have followed. For most of these medicinal species mediaeval importation seems justified, and for the more widespread perhaps an earlier Roman or Iron Age date, for their introduction into Wales.

Acknowledgements

Many people have helped provide information for this survey: my companions on the field work; Arthur Chater and Jean Green (1985) for species lists from Dinefwr and Denbigh Castles respectively; Charles Nelson who gave me data on the Irish situation. My sincere thanks to all, and especially to Philippa Tomlinson who supplied so efficiently the macrofossil records from the archaeobotanical data-base compiled at the York University Environmental Archaeology Unit.

References

Benoit, P. and Richards, M. (1963). A *Contribution to a Flora of Merioneth*, 69pp. Haverfordwest: West Wales Naturalists' Trust.

Conolly, A. (1972). A note on the flora of Welsh castles and abbeys. *Botanical Society of the Bristol Isles. Welsh Region Bulletin* **17**, 3-6.

Davis, T. A. W. (1970). *Plants of Pembrokeshire*, 82 pp. Haverfordwest: West Wales Naturalists' Trust.

Dickson, J.H. (1992). Some recent additions to the Quaternary flora of Scotland and their phytogeographical, palaeoclimatic and ethnobotanical significance. A review. *Acta Botanica Fennica* **144**, 51-57.

Garrad, L. S. (1989). Germination of long-buried seed especially *Hyoscyamus niger. Botanical Society of the British Isles News* **52**, 28.

Green, J. A. (1985). The flora of Denbigh Castle. *Botanical Society of the British Isles. Welsh Bulletin* **42**, 5-6.

Hall, A. (1986). The fossil evidence for plants in mediaeval towns. *Biologist* **33** (5), 262-267.

Hepper, F. N. (1954). Flora of Caldey Island, Pembrokeshire. *Proceedings of the Botanical Society of the British Isles* **1** (1), 21-36.

Hyde, H. A. & Wade, A. E. (1957). *Welsh Flowering Plants* (2nd ed. revised A. E. Wade), 299 pp. Cardiff: National Museum of Wales.

Johns, C. N. (1960). The Celtic monasteries of North Wales. *Caernarvonshire Historical Society Transactions* **21**, 14-43.

Jones-Pierce, T. (1963). Bardsey-A study in monastic origins. *Caernarvonshire Historical Society Transactions* **24**, 60-77.

Lacey, W. S. (1957). The flora on Ynys Seiriol (Puffin Island), with some notes of the effects of Myxomatosis. *Nature in Wales* **3**(3), 464-470.

May, R. F. (1967). *A list of the Flowering Plants and Ferns of Carmarthenshire*, 88pp. Haverfordwest: West Wales Naturalists' Trust.

Ødum, S. (1965). Germination of ancient seeds. Floristical observations and experiments with archaeologically dated soil samples. *Dansk Botanisk Arkiv* **24**(2), 1-70.

Ødum, S. (1978). *Dormant seeds in Danish ruderal soils*, 247pp. Hørsholm: Royal Veterinary and Agricultural University Hørsholm Arboretum.

Rees, F. L. (1950). *A List of Pembrokeshire Plants*, 53 pp. Tenby: Tenby Museum and West Wales Field Society.

Ridley, H. N. (1930). *The Dispersal of Plants throughout the World*, 744 pp. Ashford, Kent: Reeve.

Roberts, R. H. (1982). *The Flowering Plants and Ferns of Anglesey*, 88 pp. Cardiff: National Museum of Wales.

Saarisalo-Taubert, A. (1963). Die Flora in ihrer Beziehung zur Siedlung und Siedlungsgeschichte in den sudfinnischen Städten Porvoo, Loviisa und Hamina. *Annales Botanici Societatis Zoologicae Botanicae Fennicae 'Vanamo'* **35** (1), 1-190.

Salisbury, E. (1961). *Weeds and Aliens,* 384 pp. London: Collins.

Svensson, R. & Wigren, M. (1986). A survey of the history, biology and preservation of some retreating synanthropic plants. *Acta Universitatis Uppsaliensis* **25**, 1-73.

Svensson, R. & Wigren, M. (1990). History and biology of *Cynoglossum officinale* in Sweden. *Svensk Botanisk Tidskrift*

Synnott, D. M. (1979). *Folk-lore, legend and Irish plants.* In *Irish Gardening & Horticulture* (ed. E.. C. Nelson & A. Brady), pp. 37-43. Dublin: Royal Horticultural Society of Ireland.

Trow, A. H. (1911). *The Flora of Glamorgan,* 209 pp. Cardiff: Cardiff Naturalists' Society.

Willerding, U. (1986). *Zur Geschichte der Unkräuter Mitteleuropas,* 382 pp. Neumünster: Karl Wachholtz Verlag.

Williams, D. H. (1990). *Atlas of Cistercian Lands in Wales,* 153 pp. Cardiff: University of Wales Press.

Wynne, G. (1993). *Flora of Flintshire. The Flowering Plants and Ferns of a North Wales County,* 400 pp. Denbigh: Gee & Son.

Herbals and Other Botanical Works of the Polish Renaissance - Present State and Prospects for Research

ALICJA ZEMANEK

The Jagiellonian University Botanic Garden, ul. Kopernika 27, 31-501 Kraków (Cracow), Poland

Summary

The Polish botanical literature of the Renaissance, comprising herbals and catalogues of plants published in Polish and in Latin, is hardly known outside Poland. Publication of these works was connected with the rise of science at the University of Cracow (Kraków) founded in 1364. Between 1534 and 1613, three herbals came from the Cracow presses: *O Ziołach i mocy ich* (On herbs and their power) by Stefan Falimirz (1st edition, 1534), *Herbarz Polski* (The Polish Herbal) by Marcin of Urzędów (1595) and *Zielnik* (The Herbal) by Szymon Syreniusz (1613). Besides, lists of plants occurring in Poland (Schneeberger, 1557; Wigand, 1590; Joannicy, 1616) appeared. Polish botanical literature of the Renaissance comparable with contemporary works such as those of William Turner and John Gerard, included the first floristic records from Poland, ecological notes, information about herbal drugs and also on old Slavonic beliefs about plants. A modern study on the botanical contents of the Renaissance literature should now be undertaken, embracing determination of the plants described by the herbalist, and analysis of the beginnings of scientific botany in Poland. Interdisciplinary studies on the herbals from the point of view of the history of pharmacy, ethnology, linguistics and history of culture are also needed.

Introduction

The botanical literature of the Renaissance, comprising some hundred publications issued in many European countries, has not been investigated thoroughly. Neither the most extensive monograph by Agnes Arber (1988; 1st edition, 1912) nor the chapters in the textbooks of the history of botany (*e.g.* Meyer, 1857; Mägdefrau, 1973; Morton, 1981) and some detailed publications on this subject (*e.g.* Furmanowa *et al.*, 1959; Henrey, 1975) give a complete bibliography of old botanical books published in the 15th and 16th centuries (these books are sometimes very rare and preserved till today only as single copies). Studies on the herbals containing identifications of species described by old authors are very rare, as are studies on the beginnings of botany in particular countries. Thus, the rich botanical literature of the Polish Renaissance is hardly known outside Poland.

The publication of the first botanical books in Poland was connected with the rise of science at the University of Cracow, founded in 1364. Since about 1400, lectures on botany were provided in medical departments. The separate department of simples (medical botany) was founded not earlier than 1609.

The earliest Polish publication on the medicinal properties of herbs, *De herbarum virtutibus*, appeared in Cracow in 1532 (2nd edition - 1537) (Szymon of Łowicz 1532; 1537). To a mediaeval Latin poem on medicinal herbs, first printed in Naples in 1477, Szymon of Łowicz (born *c.* 1512), a student of Cracow University, added about 360 Polish names of plants.

Between 1532 and 1616 three herbals, three catalogues of plants and about 40 popular agricultural or medical works containing botanical chapters were published in Latin or Polish (Chmielińska, 1954).

Herbals

Stefan Falimirz - O ziołach i mocy ich (On herbs and their power) (Kraków, 1534).

The first Polish herbal, *O ziołach i mocy ich* (On herbs and their power) (Fig. 1., left) was published by Stefan Falimirz in 1534 (Falimirz, 1534). Falimirz (d. 1534), a courtier of Jan Tęczyński, the Cracow and Tarnów castellan, prepared the herbal probably by order of Florian Ungler (d. 1536), one of the first Polish publishers. The herbal is a compilation of Polish translations of Latin herbals published in Germany: *Herbarius* (Mainz, Peter Schöffer, 1484), and *Hortus sanitatis* (Mainz, Jacob Meydenbach, 1491). It contains *c.* 270 species from all over the world, described by old authors. They are mainly useful species, applied in medicine and in other ways. The plant names are written in Latin and Polish. Woodcuts of plants are taken partly from Szymon of Łowicz's publication, with some of them being drawn from nature. Three extended later editions of Falimirz's herbal appeared, under different titles, prepared by other authors (Spiczyński, 1542; 1556, Siennik, 1568) (Fig. 1.,right).

Marcin of Urzędów - Herbarz Polski (The Polish Herbal) (Kraków 1595).

In 1595 there appeared *Herbarz Polski* (The Polish Herbal), a herbal typical of the Renaissance (Marcin of Urzędów, 1595). Its author, Marcin of Urzędów [Marcin z Urzędowa], studied philosophy and medicine at the universities of Cracow (1525–1534) and Padua (1534–1538), where he took the degree of Doctor of Medicine. During his sojourn in Italy he travelled throughout the country. After coming back to Poland he became canon of Sandomierz cathedral, where he founded a garden of medicinal plants (Furmanowa *et al.*, 1959).

The herbal is a compilation of many old authors, enriched by the critical remarks of Marcin of Urzędów and his own studies on plants. He quoted 54 naturalists, philosophers, and herbalists of ancient times (*e.g.* Aristotle, Dioscorides, Pliny the Elder, Theophrastus), of the mediaeval period (*e.g.* Albertus Magnus, Mattheus Silvaticus, Mattheus Platearius), and of the Renaissance (Otto Brunfels, Leonard Fuchs). The author strongly criticized the mistakes of his predecessors (mainly those of Stefan Falimirz) in the identification of species described by the ancients and in the ways of applying herbs in treatment. For each species he gave a short morphological description, characteristics of its medicinal properties (after Galen and Dioscorides), a list of its healing virtues, and the methods of applying it in medicine, and in other ways. He also included some old legends and folk customs about plants. The herbal contains about 400 species from various geographical regions. In some cases localities of plants are included; these are the first local botanical records to be published in Poland. Polish botanical terminology, and plant names in Latin, German, and Polish are included, together with woodcuts which are partly taken from earlier Polish herbals, with many drawn from nature.

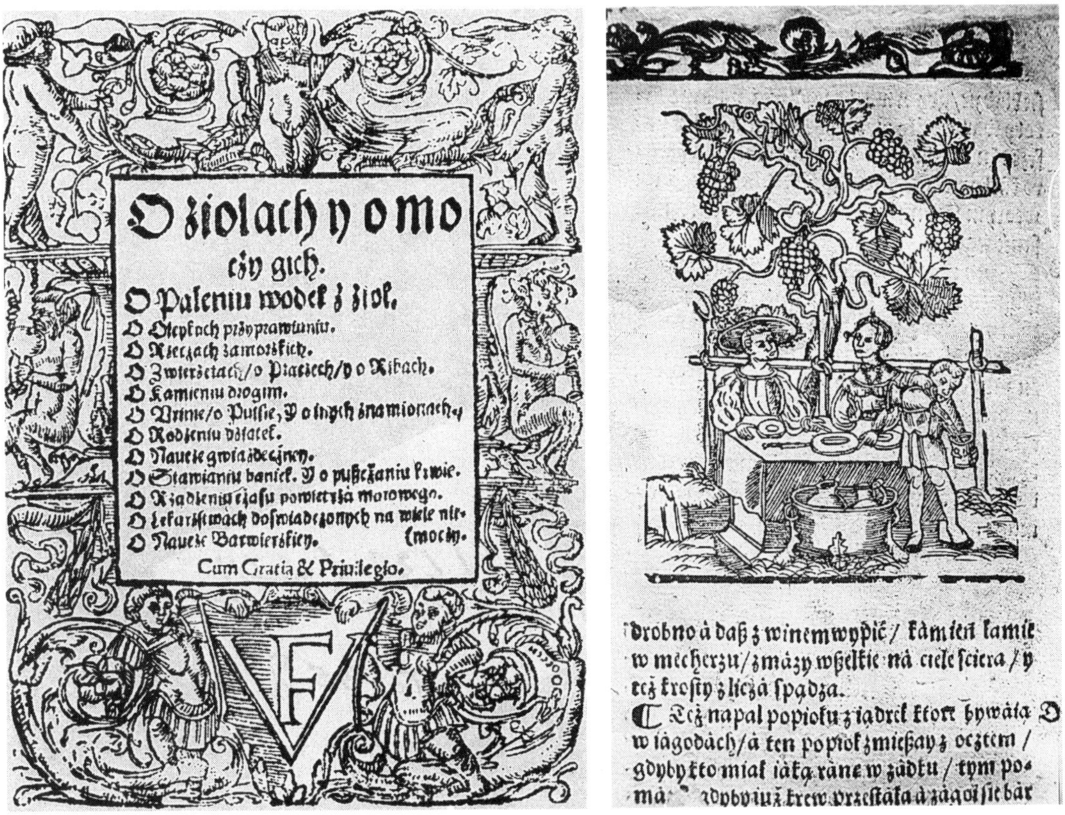

Fig. 1. *O ziołach i mocy ich* (On herbs and their power) by Stefan Falimirz. Left: Title page (Kraków 1534). Right: *Vitis vinifera* from Hieronim Spiczyński's edition (Kraków 1542).

Szymon Syreniusz - Zielnik (The Herbal) (Cracoviae 1613)

The largest work of the Polish Renaissance was *Zielnik* (The Herbal) (Fig. 2) published in 1613 (Syreniusz, 1613). Its author, Szymon Syreniusz [Syreński] (1541–1611) studied at Cracow University, where he obtained the degree of Doctor of Philosophy (1569). After travelling in Germany, Hungary, Italy, and Switzerland he studied at Padua University, where he was granted the degree of Doctor of Medicine (1577). From 1589 Syreniusz was Professor of Medicine at Cracow University.

Zielnik is a compilation of many old authors, and contains also Syreniusz's own remarks. He quoted many naturalists and philosophers of ancient times (*e.g.* Aristotle, Dioscorides, Pliny the Elder, Theophrastus), of the Middle Ages (*e.g.* Avicenna, Mattheus Silvaticus, Mattheus Platearius), and of the Renaissance (Hieronim Bock - Tragus, Otto Brunfels, Leonard Fuchs). Syreniusz consistently follows a constant scheme of description of every species. This scheme includes such elements as: morphological features of root, stem, leaves, flowers, and fruits; habitat (the first ecological records in Polish literature); time of picking for medicinal purposes; a long list of healing properties of plants; and their economic uses. Sometimes he quotes old legends and superstitions connected with plants. Some of them he criticizes, *e.g.* the use of *Artemisia vulgaris* (Fig. 3) as a magical plant against devil power. The herbal describes about 760 species from all over the world. For some species he recorded Polish localities (*e.g.* environs of Cracow, the Carpathians, etc.). Botanical terminology in Polish, and plant names in Latin, German, and Polish, are included. The Polish terminology and nomenclature of Syreniusz (taken partly from Falimirz and Marcin of Urzędów) were adopted by subsequent authors; thus they are partly preserved till today. The herbal contains about 650 woodcuts, made mainly from nature.

The herbal by Syreniusz was one of the most popular publications in Poland before 1800. The manuscripts preserved in the National Library of Vilna prove that at the beginning of the 19th century *Zielnik* was still used by Russian apothecaries.

Catalogues of Plants

Anton Schneeberger - Catalogus stirpium [...] (Cracoviae 1557)

In 1557 *Catalogus stirpium [...]* compiled by Anton Schneeberger (1530–1581) was published (Schneeberger, 1557). Its author, a Swiss born in Zurich, studied medicine and philosophy at the universities of Basel (1549–1550), Cracow (1553–1554), Montpellier (1557) and Königsberg (1560). He was a disciple of Conrad Gessner, with whom he corresponded after having settled in Poland. *Catalogus Stirpium* contains a list of the names of 432 useful plants (very often medicinal) occurring in the wild, cultivated in Poland, and exotic. Plant names are written in Latin and Polish.

Jan Wigand - Vera historia de Succino Borussico (Jenae 1590)

The next catalogue of plants, *Vera historia de Succino Borussico,* appeared in Jena in 1590 (Wigand, 1590). Its author, Jan Wigand (1523–1588) studied theology and botany at the University of Wittenberg, where he attended the lectures of Valerius Cordus. Later he lived at Miłomłyn, being a bishop of Pomerania in northern Poland. *Vera historia* includes about 440 species of plants, both occurring wild in Pomerania (the first chapter concerns the local flora of the environs of Ostróda), and cultivated.

Gabriel Joannicy - Plantarum Cracoviensium Indicem [...] (Cracoviae 1616)

In 1616 appeared *Plantarum Cracoviensium Indicem* [...] prepared by Gabriel Joannicy [Joannicius] (1565–1613), the first professor of the Department of Simples of Cracow University. All copies of this *Index,* the flora of the environs of Cracow, disappeared in the 19th century.

Scientific Value of Polish Renaissance Botanical Literature

The scientific level of botanical works published in Poland in the 16th and the beginning of the 17th centuries was varied. The earliest herbal by Falimirz, being a compilation from the old authors, was mediaeval in character, while the herbals by Marcin of Urzędów and Szymon Syreniusz were typical for the Renaissance. The catalogues of Polish plants by Anton Schneeberger and Jan Wigand display a scientific interest in the local flora and are reminiscent of similar papers published in various other countries. Similarly to other European herbalists, *e.g.* Turner or Gerard, the Polish authors made the first floristic records and ecological notes from their country, and also founded the basis of the national botanical nomenclature and

Fig. 2. *Zielnik* (The Herbal) (Cracoviae 1613) by Szymon Syreniusz. Title page.

terminology. They also recorded the folk knowledge on natural history, including many old Slavonic legends, and superstitions connected with plants.

Current State and Uses for Research of Botanical Works of the Polish Renaissance

Józef Rostafiński (1850–1928) a pioneer of the history of botany as a separate scientific discipline, carried out studies on the beginnings of botany in Poland. His

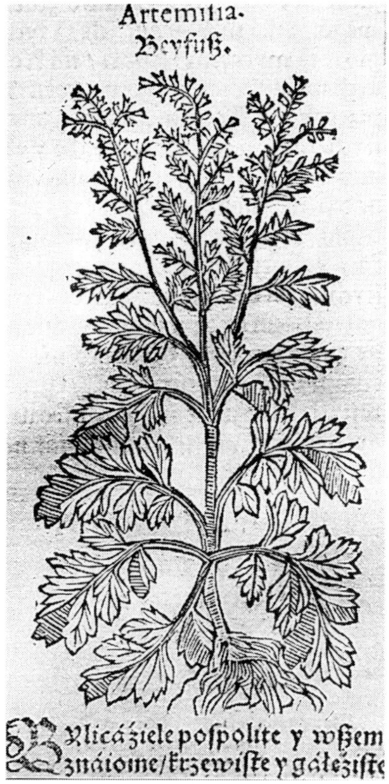

Fig. 3. *Artemisia vulgaris* from *Zielnik* (The Herbal) (Cracoviae 1613) by Szymon Syreniusz.

classical study *Średniowieczna historia naturalna* (A mediaeval natural history) (Rostafiński, 1900), containing identifications of species listed in mediaeval manuscripts, was unique on a world scale (although, unfortunately, being published in Polish, it is completely unknown to Western authors). His investigations in the old botany did not find many continuators. A modern study on the botanical contents of the Polish herbals and other works on natural history should now be undertaken. The herbals by Falimirz and Syreniusz deserve detailed studies. It must be emphasized that a synthetic monograph on the earliest botany in Poland is still lacking. Such a monograph might encompass such problems as: the level of knowledge of plant cover in Poland before 1600 (a list of the earliest floristic records, and a list of plants known in the old Poland), taxonomy, phytogeography and ecology in the works of the old authors.

Apart from detailed studies on the history of botany, a modern interdisciplinary point of view on the herbals is needed. A wide knowledge of these herbals concerns not only botany but also such branches of contemporary science as pharmacy, zoology, ethnology, linguistics, and history of culture. In Poland for over a dozen years, interdisciplinary investigations into the history of natural medicine and ethnopharmacy have been conducted. The main purpose of these researches is searching for information about plants which used to be applied in folk medicine, and which deserve to be reintroduced in treatment (Kuźnicka, ed., 1986; 1989). In spite of the abovementioned problems, the old botanical literature could be considered also from the point of view of historians and philosophers of science. Analysis of the so-called 'cultural context' of Renaissance natural history may contribute to a deeper understanding of what science is and how it developed through the ages.

References

Arber, A. (1988). *Herbals. Their origin and evolution*, 32 + 358 pp. Cambridge: Cambridge University Press.

Chmielińska, M. (1954). *Polska bibliografia zielarstwa* [Polish bibliography of pharmacy], viii + 26 + 516 pp. Warszawa: Państwowy Zakład Wydawnictw Lekarskich.

Falimirz, S. (1534). *O ziołach i mocy ich [...]* [On herbs and their power], 499 pp. Kraków: Florian Ungler.

Furmanowa, M., Michalska, Z., Parczewski, A. & Zarębska, I. (1959). Lecznictwo renesansowe na podstwie herbarza Marcina z Urzędowa [The herbal of Marcin of Urzędów]. *Studia i Materiały z Dziejów Nauki Polskiej*, Ser. B (**2**), 233-314.

Henrey, B. (1975). *British botanical and horticultural literature before 1800*, Vol. 1, 290 pp. London: Oxford University Press.

Kuźnicka, B. (ed.) (1986). *Historia leków naturalnych. History of natural drugs*, Vol. **1**, 353 pp. Warszawa: PAN - Instytut Historii Nauki, Oświaty i Techniki.

Kuźnicka, B. (ed.) (1989). *Historia leków naturalnych. History of natural drugs*, Vol. **2**, iv + 198 pp. Warszawa: PAN - Instytut Historii Nauki, Oświaty i Techniki.

Mägdefrau, K. (1973). *Geschichte der Botanik*, 314 pp. Stuttgart: G. Fischer Verlag.

Marcin of Urzędów [Marcin z Urzędowa] (1595). *Herbarz Polski* [The Polish Herbal], 6 + 488 + 22 pp. Kraków: Druk. Łazarzowa.

Meyer, E. (1857). *Geschichte der Botanik*, Vol. **4**, viii + 452 pp. Königsberg: Verlag der Gebrüder Bornträger.

Morton, A.G. (1981). *History of botanical science*, xii + 474 pp. London & San Francisco: Academic Press.

Rostafiński, J. (1900). *Średniowieczna historia naturalna* [Mediaeval natural history], Vol. **1**, xxii + 606 pp., Vol. **2**, 352 pp. Kraków: UJ.

Schneeberger, A. (1557). *Catalogus stirpium quarundam Latine et Polonice conscriptus*, 102 pp. Cracoviae: Lazarus Andreae.

Siennik, M. (1568). *Herbarz to jest ziół tutecznych, postronnych i zamorskich opisanie* [Herbal or description of indigenous, neighbouring and overseas herbs], 11 + 102 pp. Kraków: M. Szarfenberg.

Spiczyński, H. (1542). *O ziołach tutecznych i zamorskich* [On herbs indigenous and overseas], 11 + 247 pp. Kraków: Florian Ungler; 2nd ed., 1556.

Syreniusz Szymon [Syreński, Syrenius] (1613). *Zielnik* (The Herbal), 10 + 1540 + 12 pp. Cracoviae: Bazyli Skalski.

Szymon of Łowicz [Szymon z Łowicza]. (1532). *Aemilius Macer de herbarum virtutibus, cum veris figuris herbarum*, 4 + 65 + 13 pp. Cracoviae: Officina Ungleriana; 2nd ed., 1537.

Wigand, J. (1590). *Vera historia de Succino Borussico*, 11 + 153 + 5 pp. Jenae: Tobias Steinman.

Plants and People in the Isle of Man, AD 800-1800

LARCH S. GARRAD

Manx National Heritage, Manx Museum, Douglas, IOM

Summary

Aspects of the relationship between plants and people on the Isle of Man are discussed. They include the gradual felling of Man's woodlands, and the survival of pockets of the original woodland flora; plants with magical and ritual uses, especially those with calendar associations such as the 'herbs of St John' and plants associated with Tynwald Day and Lammas; herbs brought to Man by early Celtic monks; the secular herbal medicine tradition, particularly the activities of the Teare family of Ballawhane; and the herbal remedy revival which resulted from the influence of Methodism in the late 1770s. Details of herbs used to flavour beer are given as well as information on insular and imported dyestuffs.

Introduction

The Isle of Man is an island in the Irish Sea some 45 x 16 km. Uplands occupy much of the interior, reaching a maximum height of 621 m on Snaefell. Separation from the British mainland took place some 9000 years ago; it now lies about 25-27 km south of Galloway so that its most northerly point is nearer Scotland than its own southern tip. The climate is westerly and maritime ('mild, wet and windy') and has varied over the years. Thus, in early mediaeval times, as in the later Neolithic, arable cultivation was possible up to *c.* 300 m above sea level, whereas the practicable limit is now at least 120 m lower. The history of Man differed considerably from that of the adjacent islands; neither the Romans nor the Normans extended their rule to it but an independent Norse kingdom existed from 1079 to 1265, under the nominal overlordship of Norway. Subsequently it retained its own parliament, Tynwald, which has been established for at least 1000 years.

Woodland and its use

It is generally agreed that Man was over-ridden by the ice during some part of the last glaciation and thus lost its pre-existing flora. A complete forest cover subsequently developed much as elsewhere in the British Isles: oak (*Quercus robur* L. x *Q. petraea* (Mattuschka) Liebl.: cf. Allen, 1984)/hazel (*Corylus avellana* L.) on lower ground with alder (*Alnus glutinosa* L.) and willows (*Salix* L.) on wet sites and Scots pine (*Pinus sylvestris* L.) above. Birch (*Betula* L. spp.) appeared on open sites at least from the Bronze Age. The island was quite densely populated from the Mesolithic onwards and climatic changes, shading consequent on woodland development, isolation and human activity together resulted in an impoverished flora in historic times which was further depleted by woodland attrition. Archaeological evidence shows that some oak woodland and Scots pine survived as late as the beginning of the Christian era (Bersu, 1977).

It seems likely, although unproven, that from prehistory to the time of the Norse kingdom, the working of Manx minerals - copper, lead and iron - was underpinned by coppice management. There were still sufficient trees to hide an army of 300 men on Skyhill before the battle in which Godred Crovan established his supremacy over Man and the Isles about 1079. The fact that Magnus Barefoot,

King of Norway, imported Galloway timber about 1098 for his castle-building in Man is frequently adduced to prove that Man had then already lost its remaining trees. However, it may prove no more than that:

a. He had already acquired the timber there as tribute.

b. Special sizes, or seasoned timber, were required - the structures survived until *c.* 1260!

c. The Manx were so disaffected from their nominal overlord that it was unsafe for his men to go inland to fell trees.

King Harald of Man granted mineral rights to Rushen Abbey, a daughter house of Furness in the south of Man, in 1246. These were exercised by itinerant iron smelters (Garrad, 1984), while lead smelting - possibly also silver extraction - took place near their metals store at Ronaldsway (Megaw, 1960-61). Rushen Abbey still owned some woodland about 1275-1280, mainly in the northern parishes of Ballaugh and Lezayre. Interestingly, there is still a concentration in this area of woodland plants rare elsewhere in Man, *e.g.* common cow-wheat (*Melampyrum pratense* L.), three-nerved sandwort (*Moehringia trinervia* (L.) Clairv.), sanicle (*Sanicula europaea* L.) and wood speedwell (*Veronica montana* L.). Ramsons (*Allium ursinum* L.; Manx, **Craue, Garleyd Feie**), which gave its name both to the town of **Ramsey** (Old Norse) and Strooan ny **chraue** (Manx Gaelic), the boundary stream between Lezayre and Maughold, may also have been an aboriginal woodland survivor in this area. It remained rare elsewhere in Man until after 1850.

Following the end of the direct male line of the Norse kings of Man with the death of Magnus in 1265, the King of Norway ceded Man to the King of Scotland in 1266. A prolonged dispute over the island ensued between England and Scotland. This continued intermittently until King Henry IV of England granted the island to Sir John Stanley in 1405. It may be assumed that the Norse kings of Man had enforced controls on the untilled land and its resources, including woodland, but that these were disregarded during the Anglo-Scottish wars, resulting in accelerated destruction of trees and scrub. Under Sir John's heirs, the Derby lords of Man, the Lord's Forester was mainly concerned with the management of hill livestock and there was very extensive over-grazing. By 1521, an English lawyer could plausibly remark that theft of stock was not legislated against in Man because there was insufficient growth to hide stolen animals.

Repeated attempts were made to protect Man's remaining woodland. In 1629, Tynwald enacted that anyone cutting down trees other than on their own ground should be fined 10 shillings but in 1667, when S. Ratcliffe was charged with tree felling at Bishopscourt, he had to plant five, or ten, for each one he had cut down, according to whether it was his first, or second, offence. Finally, the absence of seed parents effectively prevented any major regeneration and Man became notoriously treeless.

Plants with magical and ritual use

Several plants remained in ritual, or magical, use in Man up to modern times. Some had calendar associations. On May Day Eve (**Oie Voaldyn**), a cross of rowan twigs (*Sorbus aucuparia* L.; Manx, **cuirn, kuirn**) may be placed over the inside of the door of the dwelling house. The twigs, about 5 cm long, must be broken off, not cut with iron, and are fashioned into a cross either by passing one through a slit in the other, or tying together with a wisp of unspun wool. Prior to about 1850 they were also used to protect stables and cowsheds and were also tied into animals' tails. To further avert evil, or the activities of 'the fairies', yellow flowers should be spread before the threshold. Traditionally these included primroses (*Primula vulgaris* Hudson; Manx, **Sumark**). Latterly, marsh marigolds (*Caltha palustris* L.; Manx, **Blughtyn** or **Bullught**) were more commonly used. Therefore, in Man, as in Ireland, the latter were also known as 'Mayflowers' (Manx: **Vuigh ny Boaldyn**, 'Yellow of Beltane' [the festival of 1 May]; their alternative name may be derived from Old Norse and related to the Shetland Norn **Blugda**. The Manx Folklife Survey collected many references to their importance and the need for them to be in bloom for May Eve, a cause for concern resulting from reformation of the calendar. On this date also the yellow-flowered gorse bushes (*Ulex* L. : only *U. gallii* Planch., Manx: **Aittin**, is native to Man, *U. europaeus* L. being introduced: cf. Allen, 1984) were burnt to drive out witches. The apotropaic powers of mugwort (*Artemisia vulgaris* L.; Manx, **Bollan Bane** or **Bollan Feaill' Eoin**) were also called on to protect the livestock, the plant being hung up in their shelters.

Christianity seems to have done its best to purify the plant magic associations of May Day and Midsummer Day. The vigil of St. John, 23 June (5 July after the calendar reforms), took over from Midsummer Day. St. John's Day's special plants, the French proverbial '*herbes de la St.-Jean*', included *Achillea millefolium* L., *Chrysanthemum segetum* L., *Hedera helix* L., *Hypericum perforatum* L., *Plantago major* L., *Sambucus ebulus* L., *Sedum telephium* L. and *Verbena officinalis* L., as well as *Artemisia vulgaris* (Grigson, 1975). The Manx parliament, Tynwald, meets annually in the open air at St. John's on Tynwald Day, now 5 July. This may have been an old Celtic tradition formalized on a Norse pattern when the Norse kingdom was established by Godred Crovan. It is a popular assembly, formerly for the administration of justice as well as proclamation of the law. The present structure of the meeting includes a service in the church and a procession thence to the tiered mound, Tynwald Hill.

Evidence that a Celtic midsummer public assembly pre-existed includes the rushes strewn on the processional route. A 1920s sample is mainly *Juncus articulatus* L., which probably reflects the natural growth on their source, a particular field on a nearby farm, Ballaleece, on which no Lord's Rent was payable because its holder supplied them. There is a now-vestigial tradition that, on St. John's Eve, green grass was carried to the top of Barrule 'mountain' as a tribute to Man's deity, Manannán. This ceremony was also observed on 1 [12] August, Lammas (Manx, **Laa lhunys**), or, despite the church's attempt to purify the accompanying 'rude and indecent behaviour', on the Sunday following those dates. In 1732, the curate and wardens of the parish of Lonan tried to charge offenders in the Ecclesiastical Court. Thereafter, the usual explanation of the hilltop activities was that people went to pick bilberries (*Vaccinium myrtillus* L.).

About 1925, Archdeacon John Kewley revived the custom by which those attending Tynwald wore sprigs of mugwort. This revived tradition continues. The choice of this particular 'St. John's-wort' may indicate that it was also the plant badge of the Norse kings of Man. Of the yellow St. John's-worts, *Hypericum perforatum* L. is effectively absent from Man (Allen, 1984) but another, Bog St.

John's-wort (*H. elodes* L.; Manx, **Lus y chiolgh**) has a Manx folk tale to the effect that if you trod on it after sunset on St. John's Eve a fairy horse would rise out of the ground to carry you off. Its Manx name was later garbled to 'Lusta-jough' or 'Lizzie-jog' - *jough* being Manx Gaelic for 'a drink'. *H. elodes* was also an important ingredient in herb beer.

Another seasonal celebration is to go out on Old Christmas Eve, 4 January, to see the 'myrrh' or sweet cicely (*Myrrhis odorata* (L.) Scop.) bloom at midnight. Its generic name would suggest that a similar folk belief may exist elsewhere in Europe, although apparently not in Scandinavia.

The two most important magical herbs were vervain (*Verbena officinalis* L.) and motherwort (*Leonurus cardiaca* L.). Both were known as 'Vervine', or, in Manx, **Yn Lus** (the herb of herbs), being distinguished as **He-** and **She-**, the latter probably the more elegant, *Verbena*. Both averted evil and brought good luck. Sprigs of them were put in new clothes, or tied to fishing nets. An infusion might also be sprinkled over the latter. Rituals govern the acquisition of these plants - they should be taken, not asked for and should always pass man to woman, or *vice-versa*.

Practical uses of plants

Early Christian associations

The Isle of Man had a restricted flora which therefore had to be augmented by imports to fill the practical needs of its inhabitants. Exotics brought in by people to grow frequently persist. The evidence for dates and purpose is scant but the arrival of Christianity brought fresh maritime connections with continental Europe, including the Mediterranean. A still-surviving plant probably attributable to 'Celtic' monks is elecampane, or 'horse Helen' (*Inula helenium* L.; Manx, **Lus y Rollage, Ellyn ny Compane**. Its British sites are often islands. Scottish sites include Orkney and Raasay; it is scarcely recorded from the mainland, other than coastal regions (Perring & Walters, 1982). In Ireland, it grows on Rathlin and Aran, and also on 'inland' sites which are quite often lake islets with Christian remains. It offered a cultivable alternative to the maritime native, *Eryngium maritimum* L. (sea holly). The candied roots of both soothed coughs. Its leaves were used in veterinary medicine. Reputedly (but *not* demonstrably), it was needed in dyeing wool in tartan-type blues and greens. Another early arrival may have been soapwort (*Saponaria officinalis* L.; Manx, **Brellish heabinagh**), used both in fulling and medicine.

The secular herbal medicine tradition

It has been suggested that the Druids, the priests of the pagan Celts, had a knowledge of herbal medicine which was transmitted within the secular community and then reinforced by access to continental written texts. This is supposed to be the background to such practitioners as the Welsh 'Physicians of Myddfai' (see Henderson, 1994, this volume), or the Beaton family who were taken from Ulster to Mull to serve as hereditary doctors to the Macdonalds of the Isles. Nothing is known of the Manx Druids and there are no early written documents in Manx Gaelic. However, there is evidence of some such continuity of herbal knowledge, most conspicuously among the Teare family of Ballawhane, Andreas. Oral tradition credits them with having their accumulated hereditary knowledge of the effectiveness of various locally grown herbs enhanced by the gift of a book (presumably an early translation of a classical text into a vernacular, reputedly French) from a grateful victim of shipwreck. The earliest evidence of their activities was in 1638 when Jony Tear, apparently a well known herbalist, was presented to the Ecclesiastical Court for administering strange herbs, having been '... seen together with an Irish woman in a gill pulling strange herbs', *i.e.* wild plants. She cleared herself of witchcraft and her accusers were silenced.

Rather more seriously, in 1659, Gilbert Moore and John Steon were presented for picking herbs 'to forget, to blind, to obtain property and strike people unknowingly'. Manx folklore also attributes to the Maughold witch, Berry Dhone, the claim to know of herbs that would cause a man 'to forget himself utterly'. The Teares were active throughout the 18th century; the site of one of their herb gardens beside Gat y Whing was still pointed out in the 1950s. Reputedly their plants included deadly nightshade (*Atropa belladonna* L.) but there were no relics in the 1960s. The effectiveness of their herbs may have ensured that one of their number, 'Little Daniel', survived until he was 110 in 1787. According to Feltham, writing of John Teare (Feltham, 1797), 'This gentleman's family have been long in possession of some valuable medicinal preparations which they liberally distribute for the relief of the poor'.

The influence of methodism

Interestingly, Methodist influences resulted in a revival of the use of herbal remedies. The Rev John Wesley twice visited Man, in 1777 and 1791. Wesley tried to disseminate information about apothecaries' effective dried plants and health-giving herbs which could be grown, or gathered freely. One plant brought into Man as a result may have been gladdon (*Iris foetidissima* L.; Manx, **Lus y vooin**), a diuretic. The Methodists' temperance message made little progress until the post-Revestment curb on alcohol imports but thereafter was dominant. A herbal *jough* (Manx Gaelic: drink) was the traditional alternative to beer, or buttermilk. Methodist and Rechabite influence, and perhaps also economics, helped to ensure continued use among the rural communities, particularly as a thirst quencher in the harvest fields, until at least the 1930s. Its ingredients were also considered more healthful than imported tea and coffee which were also less available to the poorer Manx after the 1765 Revestment of the insular regalities in the English Crown.

Herb beer was a sweetened decoction of herbs, fermented with added sugar and yeast but nevertheless regarded recently as non-alcoholic. Latterly it was also sold by the glass to town dwellers and holiday-makers alongside ginger and nettle beers - both containing lemons, sugar and yeast, lemonade and sarsaparilla. The core ingredients seem to have been 'a good handful' each of burdock (*Arctium minus* (L.) Bernh.), dandelion (*Taraxacum officinale* Wiggers, agg.), elegant St. John's-wort (*Hypericum pulchrum* L.), common or tree mallows (respectively *Malva sylvestris* L. and *Lavatera arborea* L., both called 'Marsh Mallow' in Man: see below), stinging nettle (*Urtica dioica* L.) and yarrow (*Achillea millefolium* L.), whence the popular 'pop' flavour of dandelion and burdock. Particular makers added other ingredients, including centaury (*Centaurium erythraea* Rafn), or the magical vervines. Wild carrot (*Daucus carota* L.) leaves might be used if the brewer lived near the sea; a dominant masculine flavour might be provided by hops (*Humulus lupulus* L.) or horehound. Both white and black horehound (respectively, *Marrubium vulgare* L. and *Ballota nigra* L.) are locally naturalized in Man, notably in the Peel area, the source of a recipe containing horehound (Manx: **Nep**).

FLAVOURINGS FOR BEERS
(* naturalized species)

*Wormwood (*Artemisia absinthium* L.; Manx, **Ullymar**).
Mugwort (*A. vulgaris* L.; Manx, **Lus Leeah**).
Ground-ivy (*Glechoma hederacea* L.; Manx, **Ard Losserey**).
*Hop (*Humulus lupulus* L.; Manx, **Lus y Lionney**).
Bog-bean (*Menyanthes trifoliata* L.; Manx, **Lubberlab**). Its Northern English dialect name is Bog Hop.
Bog-myrtle (*Myrica gale* L.; Manx, **Roddagagh**).
*Costmary (*Tanacetum balsamita* L.; Manx, **Duilagh Villish**) - naturalized quite abundantly until *c.* 1840.

INGREDIENTS FOR HERB BEER AND HERBAL INFUSIONS
(*naturalized species; + basic ingredients)

Yarrow (*Achillea millefolium* L.; Manx, **Airh Hallooin**).
+Burdock (*Arctium minus* (L.) Bernh.; Manx, **Bollan Phoa, Boagandoo**).
*+Black horehound (*Ballota nigra* L.; Manx, **Nep**).
Centaury (*Centaurium erythraea* Rafn; Manx, **Chreim Chreest** ['Christ's footsteps']).
Wild carrot (*Daucus carota* L.; Manx, **Carradje Marrey**).
+Elegant St. John's-wort (*Hypericum pulchrum* L.; Manx, **Lus y Chiolg**).
"Marsh Mallow" - *i.e.*, in Man, usually tree mallow (*Lavatera arborea* L; Manx, **Lus ny Moayll Moirrey Mooar**) or common mallow (*Malva sylvestris* L.; Manx, **Lus ny Moayll Moirrey**). Imported dried roots of true marsh mallow (*Althaea officinalis* L.) would be used in more polite households. *A. officinalis* may also have been cultivated.
* Motherwort (*Leonurus cardiaca* L.; Manx, **Yn Lus; He-vervine**).
*+White horehound (*Marrubium vulgare* L.; Manx, **Nep Vane**).
+Dandelion (*Taraxacum officinale* Wiggers, agg.; Manx, **Sharmane Muc**). Its Manx name means 'pig's sermon'. The more fashionable alternative, **Lus y Minnag** ('herb of the pinches') implies use as a cure for the blue 'pinch-marks' of scurvy.
+Hop (*Humulus lupulus* L.; Manx, **Lus Lionney**).
+Stinging nettle (*Urtica dioica* L.; Manx, **Ondaaghagh** or **Undaagagh**).
*Vervain (*Verbena officinalis* L.); Manx, **Yn Lus**; **She-vervine**).
Violet (probably *Viola odorata* L. [Manx, **Blaa Villish**], but species uncertain).

Dyestuffs

Insular

Little is known of the early history of dyeing in the Isle of Man but dyer's greenweed (*Genista tinctoria* L.; Manx, **Lus Wuigh Wooar**) is now found in a narrow zone of two of the northern parishes - Andreas and Lezayre - where it is likely to be a relic of cultivation. Interestingly, it was abundant in the Viking levels at York so its introduction as a crop may be attributed to the period of the Norse kingdom. The other 'dyer's greenweed', weld (*Reseda luteola* L.; Manx, **Wullee Wuss**), was probably grown as a dye source somewhat later, perhaps, as elsewhere, scattered among the corn since it is now widespread as a weed on disturbed ground, particularly on heaped topsoil. Yellow dyes were also obtained from the native bog myrtle (*Myrica gale* L.; Manx, **Roddagagh**) and from the old daffodil cultivar *Narcissus* 'Van Sion', known in Man as 'wild daffodil' as it is so abundantly naturalized. Its Manx name, **Lus ny Ghuiy** ('goose herb') is a reminder that it has been established so long that, like *Myrrhis odorata*, it has its own folklore. It is unlucky to pick the flower and take it into the house before the goslings have hatched, lest the goose abandon her eggs thinking the yellow flowers be her goslings. A lichen known as **Lus ny Screg** was traditionally used to dye a strong red on wool.

Imported dyestuffs

By reason of its political independence, the Isle of Man was very favourably placed to obtain the more reliable natural dyes from overseas which entered at a very low duty. Its merchants had family members resident in the eastern Mediterranean and agents elsewhere and could buy from any of the European 'India' Companies. There were also no effective constraints on the ownership of the vessels in which such goods were carried. Dyeing therefore became a fairly large-scale commercial activity at quite an early date. In the 18th century the main imports were Brazil wood (*Caesalpinia echinata* Lam.), fustic (*Chlorophora tinctoria* Gaudich.), indigo (*Indigofera tinctoria* L.), logwood (*Haematoxylon campechianum* L.), madder (*Rubia tinctorum* L.) and redwood (sanders) (*Pterocarpus santalinus* L. fil.), present in both the 1774 *Ingates and Outgates* records and the 1774 IOM Customs rates. Brazil wood yielded a red, orange or peach dye, sanders a rather fugitive red. Fustic gave a yellow dye; it was possibly later replaced by 'New', 'Zante' or 'Venice' fustic (*Cotinus coggygria* Scop.). Madder, grown in the Netherlands, was another source of red dye. Indigo (blue), was imported from America after *c.* 1750, normally in treated cakes; it was a popular present brought home by seamen after the post-Revestment rise in duty. Logwood gave a black or blue dye and is mentioned in the English *Navigation Acts* of 1660 and 1671.

References

Allen, D.E. (1984). *Flora of the Isle of Man*, 250 pp. Douglas: The Manx Museum and National Trust.

Bersu, G. (1977). *Three Iron Age Roundhouses in the IOM*. Douglas: Manx Museum.

Feltham, John (1779). *Tour through the Island of Mann in 1797 and 1798* (reprinted by the Manx Society, 1861). Douglas, IOM.

Garrad, L.S. (1984). Ballavarry, Andreas, IOM. - excavation of a medieval iron-working site. *Mediaeval Archaeology* **28**, 251-000.

Grigson, G. (1975). *The Englishman's Flora*, 542 pp. Frogmore, St. Albans: Granada Publishing Ltd (Paladin).

Megaw, B.R.S. (1960-61). Bakenaldwath and the medieval lead mines. *Journal of the Manx Museum* **6**(77), 105-000.

Perring, F.H. & Walters, S.M. (1982, reprinted 1990). *Atlas of the British Flora*, 3rd edition, 441 pp. London: Botanical Society of the British Isles.

Woodmanship

OLIVER RACKHAM

Corpus Christi College, Cambridge, CB2 1RH

Summary

Woodmanship is the craft and technology of managing and adapting wild trees and natural woodland as a resource. It is widespread in the north temperate world (as yet there is curiously little evidence for the tropics) and goes back in time to the beginnings of settled civilisation. It mostly concerns either coppice-woods or wood-pasture. The latter integrates the management of trees and wild or domesticated livestock. Wood-pasture often takes the form of savanna, and may involve pollarding or shredding the trees to yield leaves on which to feed the animals. The study of woodmanship begins by knowing the behaviour of tree species and their different responses to felling, browsing, etc. It brings together evidence from a variety of sources: written documents, archaeological timber and wood, archaeological remains on the ground, pollen analysis, ancient living trees and surviving vegetation.

Woodmanship is the craft and technology of managing wild trees and natural woodland as a resource. (**Forestry** is concerned mainly with plantations for timber production). In this paper I summarize its characteristics within the period AD 800–1800; I have published the details elsewhere (*e.g.* Rackham, 1986; 1990).

To the modern reader, the most obvious motive for woodmanship must be conservation. While conservation has certainly been a powerful motive for woodmanship in England for well over a thousand years, woodmanship goes back much earlier, to the Neolithic period, when woodland was so abundant that conservation can hardly have been much of a motive. Another motive was growing specific products such as leaves and poles. Woodmanship begins with the discovery that the stems which sprout from the stump of a felled tree, which one or two men can lift, are more useful than the original trunk. Disposing of a giant tree in wildwood, without machinery to move it, would have been as unrewarding a task then as it is now when tropical forests are grubbed out. Some big trees can be split with wedges, but this is a skilled and laborious craft, and if (like most big trees) they are crooked, spiral-grained, knotty, or fast-grown and strong the task may be impossible.

This, however, was far earlier than the period covered by this Symposium. Mediaeval north-west Europe was by no means a primitive land. The original **wild-wood** - the wholly natural vegetation, unaffected by civilization - had disappeared long ago. In England, for example, Domesday Book shows that only 15% of the country was woodland in 1086. This consisted partly of managed woods and wood-pastures derived from wildwood, and partly of new woodland that had arisen at various times on land that had previously been farmed. The abundant limes and elms of wildwood had been replaced by predominant oak, ash, and hawthorn, trees favoured by woodcutting or pasturage or that easily invade abandoned land.

Large areas of mediaeval north-west Europe had no woodland and very few trees. It was not uncommon for wooden artefacts to be transported for hundreds of km, even to places which did have trees (for example Casparie paper in this volume).

Branches of Woodmanship

Woodland

The normal method of managing woodland has been **coppicing:** felling areas of trees at intervals of between 4 and 25 years in order that they shall grow again to produce a perpetual succession of crops of rods and poles known as **underwood**. The manner of regrowth depends on the tree. Ash and maple grow from a permanent, long-lived base called a **stool**. Aspen and most elms grow from the roots. Underwood plays a very large part in buildings, fences, trackways, and firewood from the Neolithic period onward.

Timber trees are those suitable for making beams and planks. Usually they were oak if oak was available, although other species are commonly found in lower-class buildings. In the middle ages and later oaks were grown as scattered **standard** trees among underwood.

Woodmanship has probably occurred throughout Europe; as a living tradition it survives chiefly in the south, especially Italy (Moreno, Piussi & Rackham, 1982). In North America it seems to be a European introduction. As yet there is curiously little evidence for the tropics.

Wood-pasture

This combines trees with grazing animals. To some extent they are contradictory, in that the shade of the trees suppresses the pasture, and the animals are liable to eat the regrowth shoots of the trees. Various methods have been devised for compromising between the two objectives.

Wood-pasture in Great Britain originally involved cattle and sheep, but in the last thousand years has been much influenced by the husbandry of deer. This gives rise to three branches of wood-pasture:

(a) **wooded commons**, in which commoners - specific local inhabitants - have rights to the pasturage and sometimes to wood or timber produced by the trees;
(b) **deer-parks,** private land, surrounded by a deer-proof boundary on which deer are kept together with, or as an alternative to, cattle and sheep;
(c) **wooded Forests,** common-land, not surrounded by a deer-proof boundary, on which the king (or some other very important person) has the right to keep deer in addition to the commoners' and landowners' livestock.

Wood-pasture is part of the worldwide phenomenon of **savanna,** grassland (or other low vegetation) with scattered big trees. Savanna is one of the world's least understood kinds of vegetation: foresters regard it as not proper forest and agronomists as not proper agriculture, and neither party takes the trouble to understand it. Nevertheless it covers vast areas, especially in Africa. In Europe at least one-quarter of the whole of Portugal is savanna, and also large areas of Spain, Sardinia, Greece and Crete. Savanna depends in various ways on drought, fire, and browsing. In dry climates it may represent a natural transition between forest and steppe, in which tree growth is possible only if the roots can occupy more area below ground than the branches do above ground. In northern Europe wood-pasture is almost wholly a product of keeping livestock.

Hedges and non-woodland trees

Trees in hedges, standing in fields, around houses (or sites of deserted houses), on river-banks, etc. may be deliberately maintained as part of a more or less formal landscape, or may arise because farmers lack the time or inclination to prevent them from growing. Hedges between fields, or between a field and some other land-use, have often been planted, but may also arise through neglect (as they have done on a very large scale in the United States).

Woodmanship in Woodland

The earliest documents to describe specific places in the English countryside are the perambulations attached to charters of the eighth to eleventh centuries AD. These show that woodland had already much the same character as in the mediaeval and early modern period. Woods were seldom extensive; individual woodlots had names and definite boundaries; they were divided into wood-pasture and private woodland; they produced regular crops of underwood, nuts, etc.; and they were valuable land.

Individual woodlots - their extent, management, and occasionally the tree species composing them - are dealt with in detail in surveys and estate accounts from the thirteenth century onwards. Maps on a large enough scale to show details of woods begin *c.* 1580. Botanical records of a few woods go back to the seventeenth century. Many woods still in existence thus have histories traceable back to the middle ages and even the Anglo-Saxon period.

Another source of information is archaeology. The boundaries of woods are known from early maps to be very stable. On the ground they are marked by **woodbanks** - massive earthworks consisting of a bank with external ditch. These banks follow the sinuous and zigzag outlines which are characteristic of early woods. Originally they had fences or hedges: their function was to demarcate a wood-lot and to make it easier to fence against animals getting in and eating the young shoots. If the wood boundary has been altered at any time by the wood getting bigger or smaller, this corresponds to changes in the course and character of the bank. If (as often happened) the ownership of the wood has been subdivided, this results in internal woodbanks. Woodbanks thus indicate the physical and social history of a wood. The labour involved in making them reflects the great importance attached to woodland conservation; this we would hardly know from the written record, since conservation was taken for granted and seldom written down unless it failed.

Woods contain many other kinds of earthwork. Some of these, such as charcoal-hearths (level, circular platforms on which charcoal was made), result from the working of the wood. Others, such as Bronze Age funerary barrows or mediaeval ridge-and-furrow (regular undulations generated by strip-cultivation) must date from a time before the site became woodland.

Coppice stools of many species live indefinitely, and get bigger at each successive coppicing. Mediaeval woods commonly contain stools of ash, maple, wych-elm, oak, chestnut, or lime up to 3 m in diameter and several centuries old. It is sometimes possible to count annual rings in sections of the stool base.

Another independent kind of evidence is that of woodland plants. Many species, such as *Tilia cordata* Miller, *Sorbus torminalis* (L.) Crantz, *Anemone nemorosa* L., *Paris quadrifolia* L., and *Primula elatior* (L.) Hill, are strongly associated with woods known to be ancient. Others, such as *Anthriscus sylvestris* (L.) Hoffm. and *Primula veris* L. are non-woodland species which, however, may persist for centuries in newly formed woodland.

Evidence for woodland in general, though seldom for individual woods, comes from timber and underwood excavated in waterlogged sites or preserved in standing buildings. This is essential for periods when people were not writing, and is additional evidence for the historic period. For example, the written evidence for coppicing in Ireland is so meagre that some have doubted that it was ever practised, although there is good field evidence from Counties Kerry, Waterford, and Wicklow. However, recent excavations leave no doubt of the immense importance of underwood in Ireland at least as far back as the Iron Age (see, for example, O'Sullivan's paper in this volume).

In general mediaeval buildings incorporate large quantities of underwood and small timber trees. It was important to avoid the labour of subdividing the log, and therefore to use the smallest tree that would do the job. Similar traditions are found far back into prehistory (*e.g.* in the famous crannog at Flag Fen, Peterborough). The late Anglo-Saxon period is unusual for large quantities of planks cleft with wedges from very large oaks. The evidence (mainly from the London area, between the two most wooded parts of England) shows that carpenters of that period had mastered this method of making use of trees too big to move after felling, and also had access to those unusual oaks which split in a controllable way (Milne, 1992; Goodburn, this Symposium). Presumably such oaks had grown in special timberwoods – were they secondary woodland that had arisen on Roman farmland?

Domesday Book shows that coppice-woods in 1086 comprised a minority of the woodland of England – perhaps 2% of the land area; much of the rest was wood-pasture. In the next 250 years the population rapidly increased. The total woodland area declined by at least half, while the area of coppice-woods grew. Occasionally we find evidence for the rising value of woodland, and for a shift in management from accumulation of timber to the regular cropping of underwood.

Scotland

Here the woodland history is much less well known. Although Domesday Book does not cover Scotland, there can be no doubt that here, too, wildwood had long given place to something like the present landscape of farmland and moorland: the 'Great Wood of Caledon' is largely a myth of a later age (Dickson, 1992). Woodland has been less important to Scotland than to England: peat and coal were abundant alternative fuels, there was not a timber-framed building tradition (except in Edinburgh), and timber could easily be got from overseas.

Although coppicing was extensive in the eighteenth century, only a few mediaeval references are known (Anderson, 1967). Is this because it was not practised? or not recorded? or do the records not survive? or has nobody read them? Although there is some evidence for ancient wattlework (Alcock, Alcock & Driscoll, 1989), as yet Scotland lacks the extensive waterlogged excavations that have been so fruitful in England and Ireland. The answer must come from fieldwork.

Mugdock Wood, north-west of Glasgow, would be (by English standards) a typical ancient wood, with its sinuous outline, boundary bank, big ancient stools, and rich woodland flora. It also has-what is rare within ancient woods in England - peaty hollows containing a pollen record which continues to the present (Stevenson, 1990; Dickson, 1992). A number of other woods round Glasgow have field evidence for a long history of coppicing.

Scots woods tend to have a more chequered history than English and to be more difficult to interpret. Many of them are oakwoods, and have suffered attempts to grub out the stools and replace them with planted oaks. (It has even been claimed that *Quercus petræa* (Mattuschka) Liebl. does not form large, ancient coppice stools, which it commonly does in English oakwoods). The woods on the islands in Loch Lomond, though often regarded as typical ancient oakwoods, have a disappointing structure of only young, genetically-uniform trees. Some nineteenth-century owner evidently replanted them, but not with complete success: diligent search reveals remnants of the original stools on cliffs and at the shore-line.

Wood-Pasture Management

The original wood-pastures were wooded commons, mentioned in Anglo-Saxon documents as distinct from woodland. There can be little doubt that wood-pasture began long before AD 800, although it is difficult to recognize it on archaeological evidence alone.

An important difference between the wood-pastures of Britain and the Continent is that in other countries an important use of the trees was – sometimes still is – the production of leaves or bark to feed animals. This is common in countries with a short growing season, such as Norway (where dried tree-leaves supplement hay) and northern Greece (where they replace hay). In the British Isles leaves were apparently used in the Neolithic, before there was much grassland, for feeding livestock. In the historical period the practice was rare, and was more often concerned with iron rations in time of drought or great snow than with regular feeding. The English and Scottish climate is particularly suitable for grass. Animals can graze out of doors for longer than on the Continent, and the invention of the iron scythe made it possible to make hay with less effort and danger than gathering and drying tree-leaves (Rackham, 1988).

The idea of parks was introduced to England just before the Norman Conquest (AD 1066) and of Forests just after. Forests had special by-laws ostensibly concerned with protecting the deer and their habitat. To create a Forest involved declaring a defined area to be Forest, setting up a bureaucracy to administer the bye-laws, and usually introducing some deer. It did not involve any other change on the ground or in the ownership of the land. The original animals were the native red and roe deer, but soon after the Normans introduced fallow deer from southern Europe, which became the principal deer of parks and Forests.

Parks proliferated in twelfth-and thirteenth-century England. They spread to Scotland, but were much less abundant: about eighty mediaeval parks are known, scattered over the country excluding the Highlands.

Forests had their heyday in England in the twelfth century. In Scotland they are first heard of in 1110; they became more numerous, and had a much longer history, than in England. By 1500 there were, or had been some 180 Forests in the Lowlands, Southern Uplands, and south-east Highlands (Gilbert, 1979). Further Forests continued to be declared in the Highlands down to the nineteenth century. However, few of these Forests were wood-pasture. In England only about 60% of the mediaeval Forests were wooded (these contained more woodland than the rest of the countryside; few of them were entirely wooded). In Scotland the connection between Forests and woodland almost disappears: virtually all the later Forests, and many of the earlier ones, were moorland.

Wood-pastures are of two main types: **compartmented,** in which areas of trees are fenced after each felling to exclude livestock until the trees have grown again, and **uncompartmented,** in which animals have access to the trees all the time. In uncompartmented wood-pastures the trees are usually pollarded (or in former times shredded), but above the height at which livestock can reach the young shoots.

An example of a compartmented wood-pasture is Monks' Park in the Bradfield Woods, Suffolk. This was a park only briefly, in the twelfth to fourteenth centuries, and then reverted to being a normal wood like Felshamhall Wood adjacent. In its vegetation it is hardly distinguishable from Felshamhall Wood, but it still retains the compact shape of a park (for economy in expensive deer-proof fencing) and the internal earthworks of the compartmentation.

Another compartmented wood-pasture is Hatfield Forest, Essex. This contained originally seventeen coppice-woods, with woodbanks round them, covering some two-thirds of the area; the remainder was grassland and scrub with scattered pollard trees (especially hawthorns). Management continued down the centuries and has recently been revived by the National Trust, the present owners; nearly all the features are still in working order. Hatfield Forest has a complex history of conflict between the Crown (as owner of the deer), the landowners, the lords of the manor, the commoners, and the hereditary woodwards. Compartmentation operated for half of each coppice cycle and (as in many wood-pastures) did not always work well; the coppices differ from ordinary woods owing to grazing. However, on the whole the various parties held each other in balance; and the vegetation was not fragile, so that the landscape history was remarkably stable (Rackham, 1989). The 'Tragedy of the Commons', unwisely predicted by Hardin (1968), did not happen in Hatfield, or in most wood-pasture Forests and other commons. Hatfield is a microcosm of English history – it touches on Scots history as well, for the five landowners who forfeited it for English high treason included the illustrious Robert the Bruce, king of Scotland.

Uncompartmented wood-pastures, being exposed to livestock all the time, are not woodland so much as grassland or heath with trees. They tend to have browsing-resistant trees such as oak and beech (rather than ash, elm, or lime). Most woodland herbaceous plants are sensitive to browsing or trampling. Woodmanship in uncompartmented wood-pastures took the form of pollarding rather than coppicing. Some wood-pastures were sources of specially large and valuable timber trees, for example the Forest of Dean (which had a large amount of woodland) and Sherwood Forest (where the woodland was of small extent but noted as a source of very large timbers for use in cathedrals and big castles). Presumably the difficulty of replacing trees in the face of browsing caused them to be left standing for longer than in woodland.

An example of an uncompartmented wood-pasture is Staverton Park, Suffolk, which has some 4000 ancient pollard oaks as well as giant hollies and birches. Another is Epping Forest with its pollard beeches, oaks, and hornbeams. In both of these pollarding was the main management, timber being unimportant.

Hedges and Farmland Trees

Since hedges arise naturally at boundaries between fields, we would expect them to go back almost to the beginnings of agriculture. They are difficult to identify in archaeology; there are a number of probable finds in Iron Age and Roman Britain, one of which is in middle Scotland.

Hedges and hedgerow trees are abundantly recorded in Anglo-Saxon documents: as in modern Illinois or Kentucky, they were most common in the vicinity of woods, and cannot be regarded as a substitute for lack of woodland.

Hedges increased down the centuries. In the Middle Ages they were an important source of timber and underwood, even in places which also had woodland. Trees might be timber trees, pollards, or coppice stools. Species in hedges were not related to those of the local woodland; they included *Populus nigra* L., which is never recorded as a woodland tree. Although almost every parish in England had at least a few hedges, hedges and hedgerow trees were much more abundant in the peripheral counties – those which abut on the sea or in Wales – than in the Midlands where open-field cultivation prevailed. Some mountain areas, which now have stone walls, then had hedges. Hedges, however, seem not to have been common in Scotland. Numbers of hedgerow trees and trees in fields probably reached a peak in the early eighteenth century.

In the eighteenth century much of the open-field cultivation, which had covered about one-seventh of England and a similar fraction of Scotland, was abolished and made into conventional fields. This process involved the deliberate planting of tens of thousands of miles of hedges, usually with trees at intervals. At the same time farmland trees declined in the peripheral counties, both by the grubbing-out of hedges and by timber trees and pollards diminishing in the remaining hedges.

Acknowledgement

I am indebted to Dr J.H. Dickson for introducing me to Cadzow, Mugdock and other remarkable localities in the Glasgow area.

References

Alcock, L., Alcock, E.A. & Driscoll, S.T. (1989). Excavations at Dundurn, Strathearn, Perthshire. *Proceedings of the Society of Antiquaries of Scotland* **119,** 189-226.

Anderson, M.L. (1967). *A History of Scottish Forestry*. London: Nelson.

Dickson, J.H. (1992). Scottish woodlands: their ancient past and precarious present. *Botanical Journal of Scotland* **46,** 155-165.

Gilbert, J.M. (1979). *Hunting and Hunting Reserves in Medieval Scotland.* Edinburgh: John Donald.

Hardin, G. (1968). The tragedy of the commons. *Science* **162,** 1243-1248.

Milne, G. (1992). *Timber Building Techniques in London* c. *900-1400*. London and Middlesex Archaeological Society.

Moreno, D., Piussi, P. & Rackham, O. (ed). (1982). Boschi: Storia e Archeologia. *Quaderni Storici* **49,** 7-163.

Rackham, O. (1986). *The History of the Countryside*. London: Dent.

Rackham, O. (1988). Trees and woodland in a crowded landscape – the cultural landscape of the British Isles. In *The Cultural Landscape – Past, Present and Future* (ed. H.H. Birks *et al.*), pp. 53-78. Cambridge: Cambridge University Press.

Rackham, O. (1989). *The Last Forest: the Story of Hatfield Forest.* London: Dent.

Rackham, O. (1990). *Trees & Woodland in the British Landscape.* 2nd edition. London: Dent.

Stevenson, J.F. (1990). How ancient is the woodland of Mugdock? *Scottish Forestry* **44,** 161-172.

Trees Underground: New Insights into Trees and Woodmanship in South-east England *c.* AD 800-1300

D.M. GOODBURN

Museum of London Archaeology Service, Number One, London Wall, London, EC2Y 5EA

Summary

Study of wood recovered from waterlogged archaeological sites in London reveals much about the trees that grew in the city's hinterland between about AD 800 and 1300. Tentative reconstructions of some specific types of trees and 'treeland' have been made by combining three-dimensional recording of the timbers and roundwood with examination of how the wood was used, tree-ring analysis, species identification and experimental work in ancient woodland. It has also been possible to gain some appreciation of general trends during the mediaeval period such as the intensification of woodland management and the disappearance of wildwood conditions.

Introduction

This paper is dedicated to the memory of the late Helen Hibberd whose work on botanical aspects of early woodmanship has been greatly valued by many of us. What follows is a brief summary of the approach used by the Museum of London Archaeology Service to investigate early historic trees and treeland in the London area. The term 'treeland' is used to describe any land on which trees grow, including wildwood, high forest, coppice with standards woodland and orchards. Our approach to reconstructing past treeland has been inspired by the pioneering work of Oliver Rackham on standing mediaeval timber buildings from late mediaeval times (Rackham, 1972; 1976).

However, for earlier periods we must look to underground evidence. London is fortunate to have many waterlogged zones in which large amounts of timber and roundwood have been found well preserved from the Bronze age to post-mediaeval periods. Structural remains of many types have survived, including buildings, wharfs, bridges, boats, wells, wattle fences, trackways and groups of reused timbers. The often hurried circumstances of rescue archaeological work in London are not ideal, but where possible we try to carry out the following procedure when excavating waterlogged timber and roundwood (*e.g.* Spence, 1990; Goodburn, 1991).

1. Timber and roundwood structures are first exposed, recorded in plan and the best preserved sections then drawn in elevation and photographed.
2. These priority areas are dismantled and individual components drawn, described on pro-forma sheets and selectively photographed. We pay particular attention to the following features; size, shape, type of conversion (section of tree used), method of conversion (*e.g.* by sawing, by cleaving) and the major features of tree anatomy, knots, grain, presence of sapwood or bark. Finally, evidence of reuse is noted.
3. Each timber is then sampled for tree-ring study; structures of small roundwood are also systematically sampled.

4. Given sufficiently well preserved material, attempts at graphic reconstruction of the following will be made, drawing on the results of the tree-ring and species identification work and the drawn field record; the parent log, the parent tree, and where the evidence is particularly strong the parent treescape.

For a solid basis to this work we have been carrying out experimental work in ancient woodland on the edge of London for the last 6 years. Reconstructions of ancient structural woodwork have been made using period type tools and trees of the closest possible characteristics to those used originally (*e.g.* Goodburn & Redknap 1988; Goodburn 1992).

Close collaboration is possible in London between tree-ring specialists and field workers who all work 'in house'. The following section of this paper jointly summarizes aspects of the work of this author, and tree-ring specialists I. Tyers, N. Nayling and H. Hibberd. Tree-ring work is used here as an umbrella term for species identification, ring-counting, and for samples with 50 or more rings, ring measuring. If ring sequences can be cross-matched to standard regional reference curves accurate dating and provenancing timber to broad geographical zones is possible.

The archaeological woodwork from London is dominated by oak (*Quercus robur* L. or *Q. petrea* Liebl.). However, the form, age and type of the oaks used varied greatly, principally as a response to varying methods and intensities of management. Three worked examples are presented.

Wildwood or high-forest oaks

Timbers cleft and hewn from very large, straight grained oaks are particularly common on London sites in the 10th to 12th centuries, becoming much less common later. The trees were often worked into thin wide radial boards for building, boat and cooperage use. Typically these great parent oaks were 2-300 years old, between about 0.7 and 1.2 m in diameter and very slow grown (Fig. 1a). Such oaks will grow only in conditions of high dense woodland which must either have been wildwood or perhaps very overgrown Roman managed woodland. Oak trees with these characteristics are extinct in Britain today. Such material is found alongside different types of timber and small roundwood showing the contemporary treeland was a complex mosaic of types.

Pollarded oaks

The woodsmanship practice of 'pollarding' where a tree is cut between about 1 and 4 m above the ground and the rapid regrowth repeatedly harvested is well known from later mediaeval times (Rackham, 1976:22). However, recent work in London can now take the dating of the practice back to about AD 1180, supporting theories of early oak pollarding based on charter evidence (Rackham, 1976:54).

A very high proportion of mediaeval sawn oak planks are of very short lengths 1.6-3.5 m long but are often rather wide, *c.* 0.5-0.8 m. Typically the logs they were cut from had wide rings and rather curved knotty grain, quite different to the material used for most cleft boards. The method of conversion was by see-sawing, introduced about 1180 AD (Goodburn 1992:113). Experimental work has shown that this sawing method was particularly well adapted to the use of the short thick trunks of large pollards (Fig. 1b). Detailed analysis of the ring sequences in several recently excavated 13th century oak planks shows cessations of summer growth every 13-16 years for 60 years or so, in a sequence of otherwise rather wide rings (Fig. 2). This patterning is similar to that in modern pollards and together with the morphology of the planks it is taken as evidence of pollarding rather than the related practice of shredding. Shredding provides longer but even more knotty timber. The planks were in a structure with other timbers which were felled in 1241; as none of the timber showed signs of reuse it is likely that the parent pollard was felled about the same time. Therefore this particular pollarding began in the 1180's. It is now clear why the matter of the ownership of pollard tree bollings was considered important even though they would generally have little value to a modern forester.

Evidence for coppice management

Coppicing is a well known woodmanship practice by which trees are repeatedly cut down to ground level and the vigorous regrowth harvested (Fig. 1c). However, many of the details of this practice in early periods are little known. Roundwood structures of early mediaeval date such as wattle pit linings, hurdles and fences are relatively common finds in London. The weaves and types of rods used for the

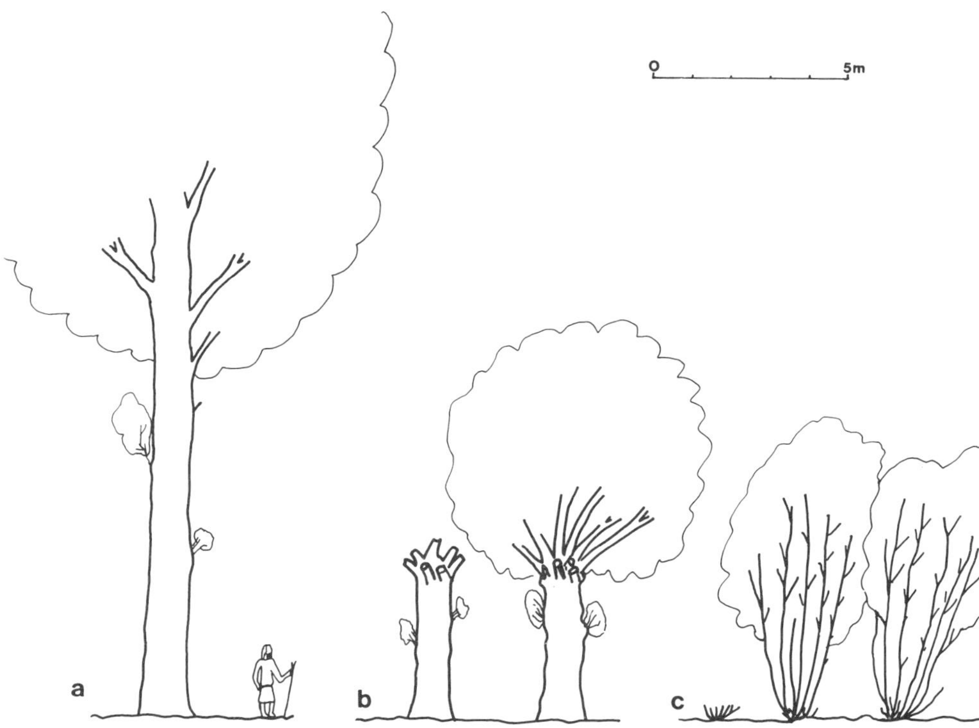

Fig. 1. Hypothetical reconstructions of medieval parent trees for; a) A typical wildwood oak 2-300 years old used for the production of radially cleft boards 9th-12th*c*. b) A pollard oak of about AD1240 used to make short, wide see-sawn planks. c) Young oak and hazel coppice less than 7 years old such as was used for wattle hurdle production in the London area *c.* AD1200 .

structures varied considerably. Some structures have a 'home-made' appearance with irregular rods of varied species and age, others are very neatly woven with long regular rods such as can be produced only by pollarding or coppicing.

Pilot analysis of this varied material has shown some interesting patterns, with some structures clearly being made from casually collected local hedgerow or orchard thinnings and others being produced from systematically coppiced woodland. For fine wattle work such as hurdles the rods used were small and whole rather than cleft as in recent practice. This suggests that they were from coppice cut on a short rotation. Detailed tree-ring analysis shows this to have been the case for some finely made structures such as a hurdle dated to *c.* AD1200. Here the uprights were 6 year old oak and the rods a mix of 6 year old hazel (*Corylus avellana* L.) and 3 year old oak.

Conclusions

This short paper has attempted to demonstrate the potential of waterlogged archaeological woodwork to increase our understanding of aspects of early woodmanship. It is to be hoped that charcoal identification, woodland pollen analysis and historical, ecological and archaeological studies of ancient treeland in the London area will refine and advance this vital area of historical ecology.

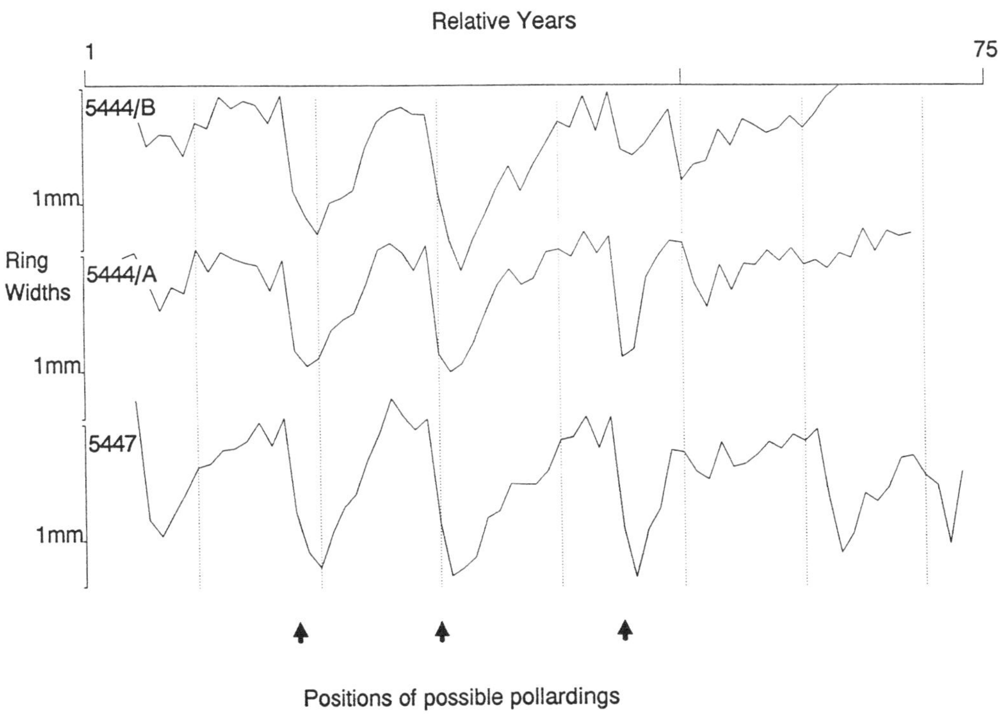

Fig. 2. Three essentially identical tree-ring sequence graphs for samples taken from three see-sawn oak planks from the same structure dating to about AD1241. The repeated growth halts are indicative of pollarding starting in the 1180's.

Acknowledgements

This work would have been impossible without the dedication of the Museum of London excavation and specialist staff, particularly I. Tyers, H. Hibberd and N. Nayling. I. Tyers kindly commented on a draft of this paper, but any remaining errors are the author's responsibility.

References

Goodburn, D. & Redknap, M. (1988). Replicas and wrecks from the Thames area. *London Archaeologist* **6**(1), 7-10, and 19-22.

Goodburn, D. (1991). Waterlogged wood and timber as archives of ancient landscapes. In *Wet Site Excavation and Survey* (ed. J. Coles & D. Goodburn), pp. 51-53. Museum of London.

Goodburn, D. (1992). Woods and woodland; carpenters and carpentry. In *Timber Building Techniques in London c. 900-1400* (ed. G. Milne), pp. 106-130. London and Middlesex Archaeology Society Special Paper 15.

Rackham, O. (1972). Grundle House: the question of timber in certain East Anglian buildings in relation to local supplies. *Vernacular Architecture* **3**, 3-8.

Rackham, O. (1976). *Trees and Woodland in the British Landscape,* 204 pp. London: Dent.

Spence, C. [ed.] (1990). *Museum of London Excavation Manual,* 2nd edition. Museum of London.

Some Examples of the Use of Wood in the Historical Town of Groningen (The Netherlands)

W.A. CASPARIE

Biologisch-Archaeologisch Instituut, Rijksuniversiteit Groningen, Poststraat 6, 9712 ER Groningen, The Netherlands

Summary

Many wooden artifacts have been excavated from deposits formed in Groningen from the 7th to 18th centuries AD. Identifications have been made of wood used for building, for coffins, cups, spindles, a sun watch, a bird cage, a doll and other objects. Inferences are made about woodmanship, gardening and trade.

Introduction

A number of wooden artifacts and their functions will be discussed, all originating from excavations in the historical subsoil of the town of Groningen. The town is situated in the north-eastern part of the Netherlands, on the northern tip of an upland area, south of the marine clay district and west of extensive raised bogs. Within the area of the mediaeval town, many excavations have taken place, especially in the last 15 years. The wood remains from a number of these have been investigated.

The Material

Most of the wood originated from the fill of cesspits, refuse pits, wells and ditches. In some excavations the fill of these pits was sieved. In these cases, also, very small wood remains could be retrieved. Up till now more than 10,000 pieces of wood have been sampled and investigated, mostly collected in large-scale excavations from 1987 on. This has resulted in a record of more than 50 different wood species for historical Groningen. A few examples of the use of wood in the historical period, which can inform us about woodmanship, wood production, craft and trade, will be presented here. Most of the wood derives from four major excavations: inside St. Martin's church, the nearby Oude Politiebureau-Martinikerkhof, the Wolters-Noordhoff Complex and the Casino (Fig. 1).

Results

Cemetery with coffins of hollowed-out oak trunks

In the funerary ritual in the Early Middle Ages, inhumations and cremations occur side by side. In historical Groningen part of a cemetery, which mostly contained coffins made of hollowed-out thick tree trunks of *Quercus* (oak), was excavated underneath St. Martin's church, the main church of the town, with several mediaeval building phases. More than 30 coffins were excavated, the radiocarbon dates of which indicate that the cemetery was in use from the 7th to 10th centuries AD (Lanting, 1990: 155-174). A schematic view of such a coffin is presented in Fig. 2.

The diameter of the oak trunks used for these inhumations was 70-75 cm; the trees had mostly reached an age of 100-150 years. They were remarkably fast-growing oak trees, deviating clearly in size and growth characteristics from oaks selected for structural timber and other building wood. For the latter purposes in

Fig. 1. Central part of the town of Groningen with some large-scale excavations. 1. St. Martin's church, 2. Oude Politiebureau-Martinikerkhof, 3. Wolters-Noordhoff Complex, 4. Casino. Scale 1:15,000.

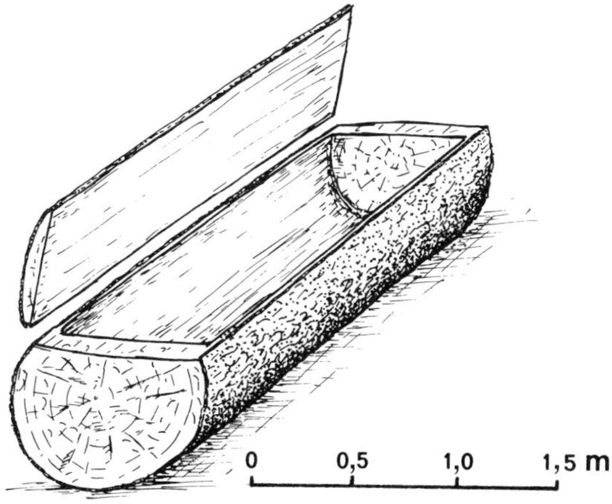

Fig. 2. Schematic representation of a coffin made of a hollowed-out oak trunk. St. Martin's church.

general more slowly-grown oaks with a diameter of 40-45 cm were selected, which originated from comparatively dense forest stands, where fast-growing trees are lacking. The faster-growing, thicker oak trees, like those used for the coffins, did not occur in Dutch forests, which grew under more or less natural conditions. Given their lavish growth, it is very likely that these oaks have been cultivated in one way or another. This must have taken place on very fertile and stable soils, so that after 100-150 years the thick-grown oaks could be felled. Such rapid growth rates will occur only in more or less free-standing trees, tended carefully over 100 years (Casparie, 1991: 106-107); this is evidence of forest management based on skilful woodmanship.

Tree trunks for this type of coffin cannot be produced instantly. The long time of production of 100-150 years and the spatial requirement of free-standing trees with their large canopy, very clearly implies long-term forestry planning, for the purpose of the funerary ritual, from the 7th to the 10th century. It can be calculated that the spatial need for such wood production concerns at least 1650 ha in the north-eastern Netherlands.

Stave cups

Well known from many cities of the European continent, but also from London, are small stave cups. Five types can be distinguished, according to the number of hoops fixing the staves. All five types are known from Groningen (Casparie & Helfrich, 1993: 434-443). Types 1 and 2 (Fig. 3) generally date from the 12th to 14th

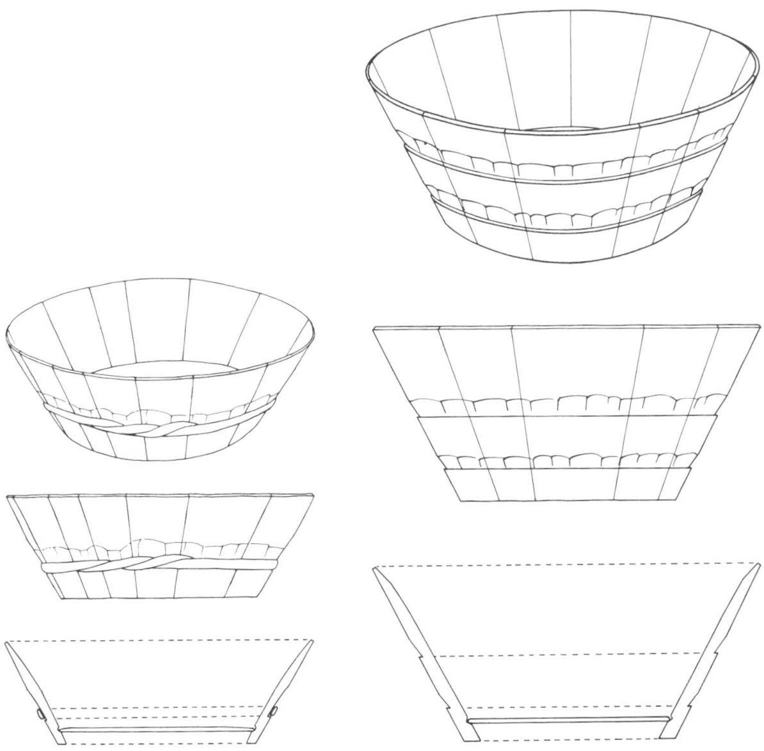

Fig. 3 . Stave cups, types 1 and 2. *Picea* and *Frangula alnus*. Wolters-Noordhoff Complex. Scale 1:3.

centuries; types 3-5 from the 14th to 16th centuries. The staves and the round bottoms are of *Picea* (spruce). The hoops are generally made of thin split twigs of *Salix* (willow). The best-known find-spot of these cups of types 1 and 2 is Lübeck in northern Germany, one of the mediaeval ports on the Baltic. Hundreds of these cups were found there, in a cellar, evidently a trader's stock (Neugebauer, 1975: 117-137).

One single cesspit in Groningen produced a large concentration of more than 50 cups each with a single hoop, together with about 10 cups each with two hoops, many of them still in the original assembled form. Staves and bottoms were manufactured of spruce. The hoops were made of split twigs of *Frangula alnus* Mill. (alder buckthorn).

The Groningen cups were clearly imports, for *Picea abies* (L.) Karsten (spruce) is not native in the Netherlands, as is shown in Fig. 4, which depicts the natural distribution area of spruce in mediaeval Europe. Hence it is likely that these cups were imported from the Baltic area, testifying the importance of Hanseatic trade. Trade with the Upper Rhine region in central and southern Europe was minimal in the 13th and 14th centuries.

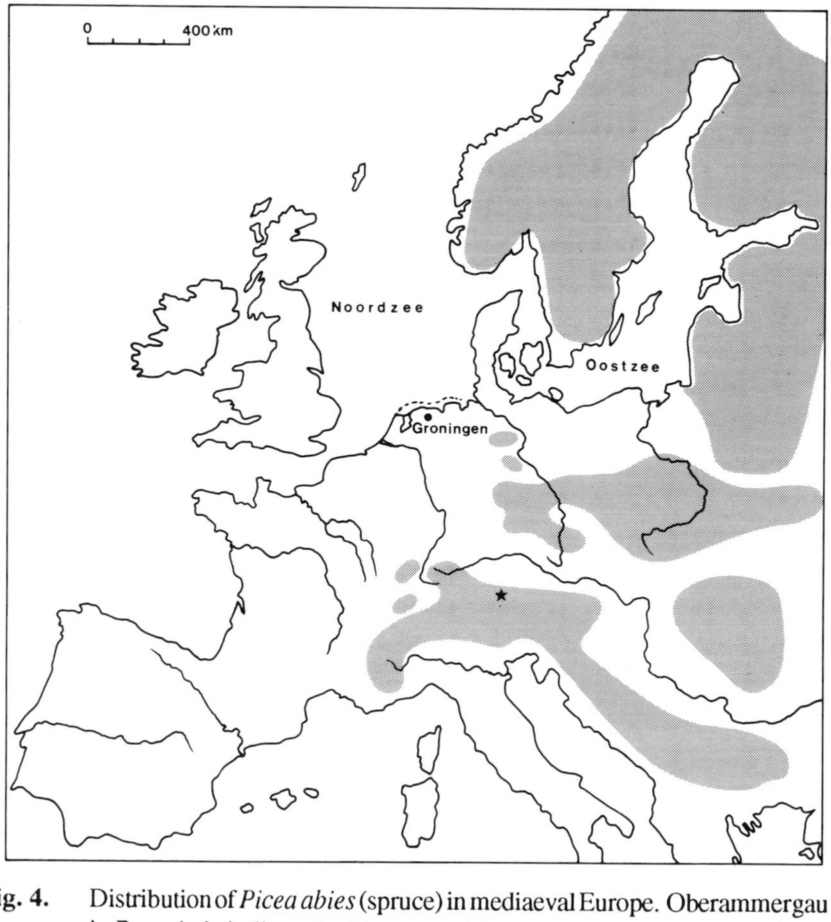

Fig. 4. Distribution of *Picea abies* (spruce) in mediaeval Europe. Oberammergau in Bavaria is indicated with an asterisk.

The use of twigs of *F. alnus* as hoops is up till now without parallels. This batch of cups very likely represents a one-off importation by a single trader. The cups were very definitely not luxury goods; they were simple and probably cheap tableware.

Sunwatch

The sunwatch of *Buxus* wood (boxwood) (Fig. 5) dated to about 1500 was probably a real luxury item. Remnants of the copper pin for the compass needle and of the copper case are still present in the wood. The wood of which the sunwatch is made definitely indicates import, perhaps from the Mediteranean area. It is still unknown for which meridian the watch was made.

The most striking difference between this sunwatch and those found in the wrecked warship Mary Rose, which are somewhat older than 1535, is the eccentric place of the compass of the Mary Rose sunwatches. This suggests that the Groningen sunwatch is a somewhat less sophisticated one and possibly older. This agrees with the archaeological dating of about 1500. For the Mary Rose sunwatches it has been suggested that they were manufactured from wood from southern Germany.

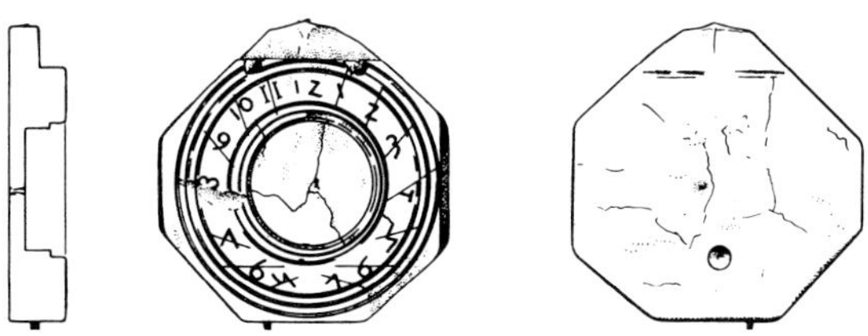

Fig. 5. Sunwatch. *Buxus*. Oude Politiebureau-Martinikerkhof. Scale 1:1.5.

Small utensils of Ilex wood

In various sites of Groningen small artifacts made of wood of *Ilex* (holly) have been found in cesspits and other pits. They are roughly contemporary, dated between 1560 and 1600 (Casparie, 1988: 132-140). The wood species has been recorded 20 times so far. Seven of these items were found to be turned spindles, and more or less identical (Fig. 6). The others are small artifacts of unknown function, also turned and partly carved.

Ilex is native in forests in the Netherlands, though not in large numbers. The tree has been planted in gardens since the 18th century. The wood is rather hard and brittle when worked. It is not very durable and generally not highly valued. International or even regional trade of utensils is unlikely. In the case of the Groningen finds, which all can be dated within a rather short period at the end of the 16th century, it may be assumed that a local craftsman had this wood at his disposal. Apart from holly, he may also have used wood of *Cornus* (cornel) and *Sambucus* (elder). However, none of these wood species were often used for turned woodworking.

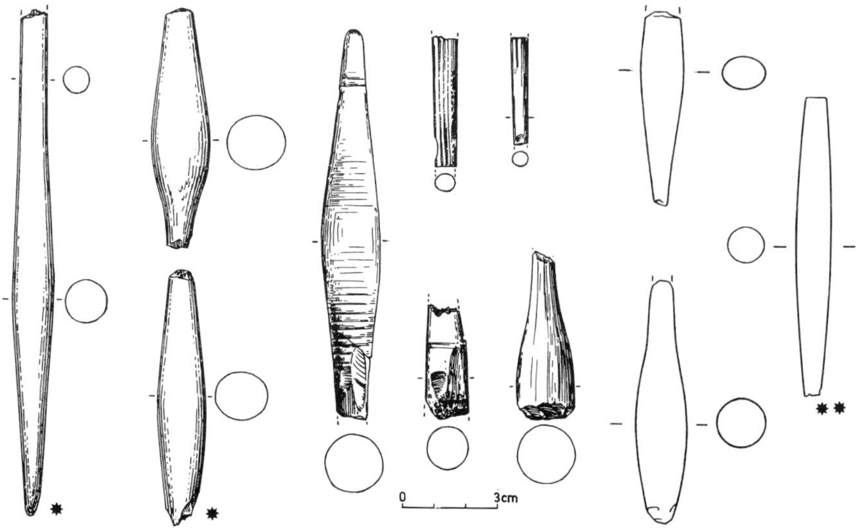

Fig. 6. Artifacts of *Ilex* wood and also of *Sambucus* (*) and *Cornus* (**). Oude Politiebureau-Martinikerkhof, Wolters-Noordhoff Complex and Casino. Scale 1:3.

Pinus and Picea as building wood

In the 16th century, coniferous wood of *Pinus* (pine) and *Picea* (spruce) appeared as building timber, as is demonstrated by the wood spectra. The beams and boards of such wood are definitely imports from Scandinavia, as is also historically documented. They were transported in bulk. Here an example illustrates how this coniferous wood replaced oak-wood as the most important building timber.

In the south-eastern part of mediaeval Groningen, the infirmary of St. Gertrude's Hospital was built about 1405. An excavation at this site uncovered a cesspit, dated to about 1570/1580. This cesspit contained many remains of building timber, such as fragments of beams and boards, especially of oak, pine and spruce. This suggests that the cesspit must have been open for some time during a repair of the infirmary. Most of the oak timber remnants were rotten, decayed and weathered; very likely old demolition wood. Most of the pine wood was not decayed. On the contrary, it looked like fresh wood with clear sawing traces and axe marks still visible and therefore it can be interpreted as new coniferous building timber. The original building with brick walls in the style of the 15th century must have had a roof and interior of mainly oak wood. Wood of *Pinus* and to a minor extent of *Picea* was employed for the repairs, during which the cesspit must have been open for one reason or another and fragments of the demolition and repair timber could easily drop in (Casparie, 1988: 132-140). Which building was repaired can be reconstructed from some old maps of the city. The bird's eye view of 1565 (Fig. 7) shows the south-eastern part of the town. The black arrow indicates the infirmary in question, to which the excavated cesspit belonged. In a bird's eye view of 1575, the same building is depicted without a roof-turret. The conclusion can be drawn that a few years after 1565 the oak wood parts of the building were repaired with coniferous wood, especially pine, and that the turret was entirely removed.

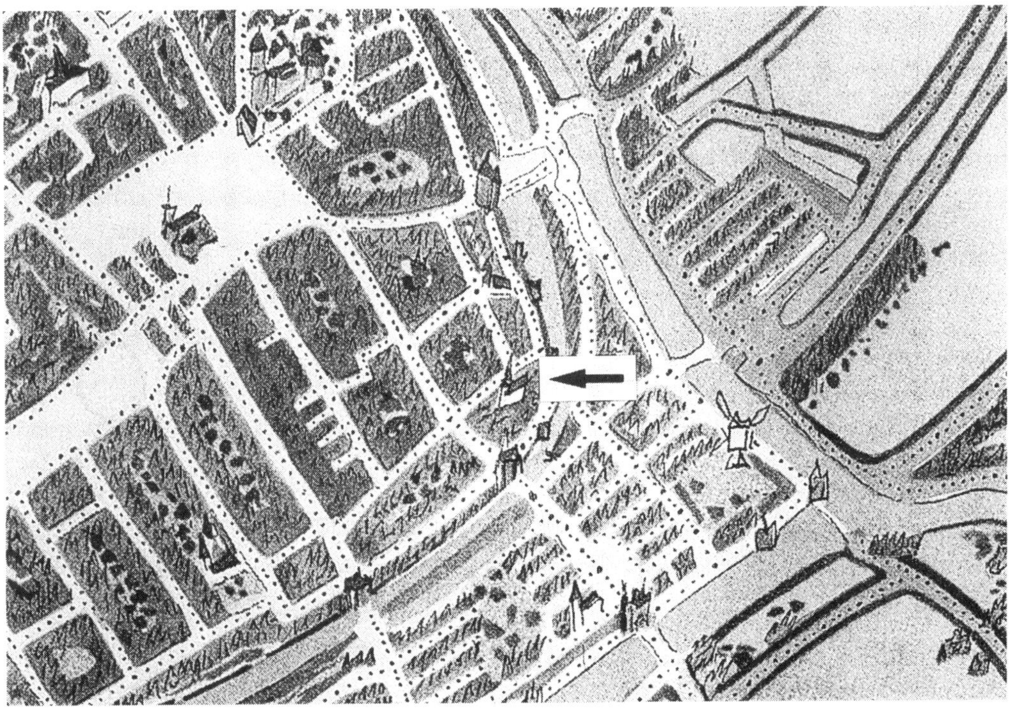

Fig. 7. Detail of a map of the city of Groningen, dated to *c.* 1565. The infirmary of St. Gertrude's Hospital with its roof-turret is indicated with a black arrow.

16th and 17th century gardens in the town centre

By sieving the fill of cesspits, numerous small pieces of wood have been retrieved, including chips and parts of twigs and branches, of which often the bark was still present. Part of this wood derived from trees and shrubs of nearby gardens. Some of these twigs may be the remains of wattlework, brooms, barrel-hoops and so on. It is possible to eliminate those kinds of wood which generally represent artifacts or construction wood. What remains after this elimination are the twigs and branches of the local trees, giving insight into what was grown in the surrounding gardens.

Table 1. Oude Politiebureau-Martinikerkhof. Garden trees and shrubs.

Trees and shrubs	Fruit trees	Ornamental shrubs
Quercus	* *Malus-type*	* *Ilex aquifolia*
Fagus	*Prunus* 2-3, *avium*-type	*Viscum album*
Ulmus	*Prunus* 2-4, *padus*-type	* *Cornus sanguinea*
Fraxinus	*Prunus* 3-5	*Hedera helix*
Acer campestre-type	*Prunus* 4-6, *spinosa*-type	*Ligustrum*
Alnus	* *Prunus* 5-7, *domestica*-type	* *Viburnum*
Betula	* *Prunus* 6-8, *persica*-type	* *Lonicera*
Corylus	* *Sambucus racemosa*	* cf. *Daphne*
* cf. *Carpinus*	*Vitis*	* *Leguminosae*
Salix	* *Ribes rubrum*	*Rhamnus*
		* *Berberis*
		* *Taxus*
		Juniperus

In Table 1 the garden wood species are listed from a number of 15th-, 16th- and 17th- century cesspits in the centre of the historical town, situated close to each other. The species can be divided into three groups: trees, whose wood is also known from artifacts, fruit trees, and ornamental trees, mostly shrubs. An asterisk indicates those species, that appear for the first time in the second half of the 16th century or in the 17th century. There is a remarkable increase in ornamental shrubs, such as *Ilex, Cornus sanguinea* L., *Viburnum, Ligustrum, Lonicera,* perhaps *Daphne,* a Leguminosae species, *Berberis* and *Taxus.* New fruit trees are also recorded: plum-type and peach-type, *Sambucus racemosa* L. (red-berried elder) and *Ribes rubrum* L. (red currant-type). It suggests that in the course of the 16th century especially pleasure-gardens came into existence, in which fruit trees still remained important, with an increase of imported fruit-tree species.

Bird cage

The remains of the bird cage (Fig. 8) can be dated to the 17th century. The frame is made of *Quercus* wood, and the thin bars of carved *Buxus* (boxwood). The distance between the bars is about 1 cm, which means that the cage could be used for small birds such as finches. This cage apparently had a horizontal setting of the bars. In 18th- and 19th-century cages the bars are almost always vertically placed. The cage is an imported artifact, perhaps originating from the Mediterranean.

Stoppers for wine bottles

In a refuse pit, dated to the end of the 18th century - perhaps about 1790 - more than 100 stoppers used for wine bottles came to light (Fig. 9). The pit also contained a large number of wine-bottles of the 'lady's leg' type.

The stoppers are coated with a 1 to 2 mm thick very hard, dark substance, perhaps a kind of pitch, and are up to about 5 cm long. They were not made of cork, but of an organic, non-woody material, in which the cell structure resembling parenchymatic tissue is still visible. Given the size of the stoppers, they could not have been made of a cereal. Perhaps they were made of potato tubers, but it is very difficult to prove.

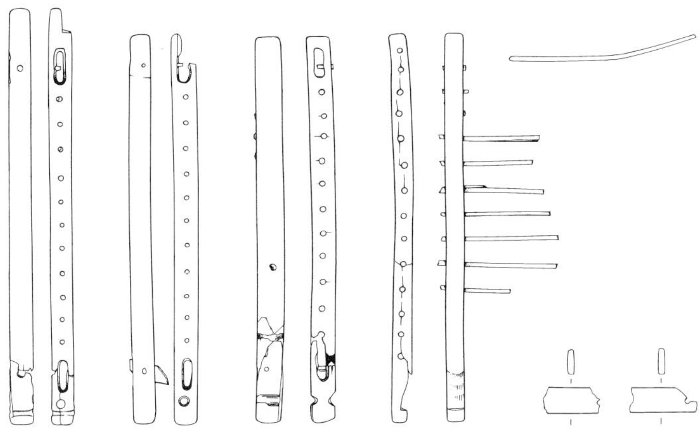

Fig. 8. Bird cage. *Quercus* and *Buxus*. Oude Politiebureau-Martinikhof. Scale 1:4

As wine was not produced in the Netherlands, we must be dealing with an import, perhaps from France or central or southern Germany.

Wooden doll

From the same refuse pit that produced the stoppers and the wine bottles, a wooden doll with a length of 15.5 cm (Fig. 10) emerged, which is also dated to about 1790 (Casparie & Helfrich, 1993: 227-235). The arms and legs can be moved, but the moving mechanism was not preserved. It must belong to the type shown in Fig. 11. The doll tumbles like an acrobat, when the mechanism is used.

Body and arms are made of *Populus* (poplar); the legs, the pivots and the small pins in the ends of the pivots are of *Picea* (spruce). This suggests an import from a craft centre situated in the area where *Picea abies* is indigenous (Fig. 4). On stylistic grounds it became clear that this doll had been manufactured in the craft centre of Oberammergau in southern Bavaria by the firm of Georg Lang Seliger Erben, a wholesale merchant and distributor of woodcraft. This firm, founded in 1775, still exists as a flourishing large woodcraft workshop. Since the late 18th century, firms of this kind had their agencies over the whole of Europe, as well as in Africa and America.

Details of the trade connection to Groningen, covering a distance of about 800 kilometres, are not known.

Epilogue

The Groningen archaeological wood produces quite a good picture of the historical use of wood. Forest management was practised already in early mediaeval times. Trade in wood, especially wooden products, occurred over large distances. In contrast, local wood from the town of Groningen and its surroundings was also used, as is illustrated by an excavated wooden shoe, 'Klomp', made of *Populus* (poplar) wood from the Wolters-Noordhoff complex (Fig. 12). This is a typical Dutch product from the 19th Century; wooden shoes of this type were manufactured in the vicinity of historical Groningen.

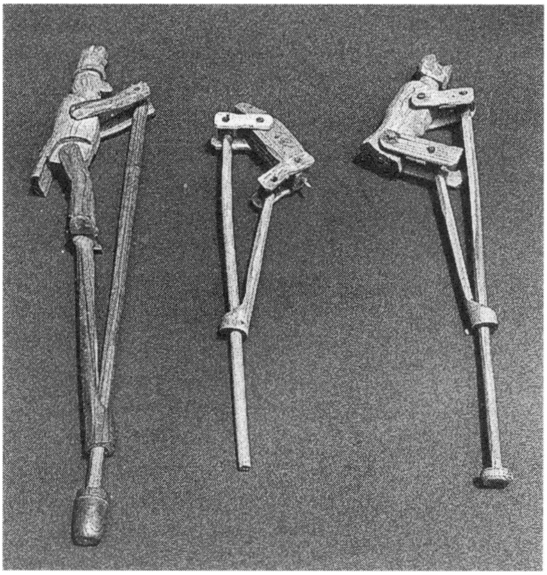

Fig. 9. Stoppers for wine bottles made of parenchymatic tissue, perhaps of potato tubers, with a hard coating. Length of the stoppers: 4-5 cm. Wolters-Noordhoff Complex.

Fig.10. Wooden doll 'acrobat', length of the doll: 15.5 cm. *Picea and Populus.* Wolters-Noordhoff Complex.

Fig. 11. Some acrobat-dolls of the famous collection of wooden toys from Grödnertal, northern Italy, property of Mrs Rita Stäblein at Icking-Irschenhausen, Bavaria.

Fig. 12. Wooden shoe; 'klomp', length: 33 cm. *Populus. Wolters-Noordhoff Complex.*

Acknowledgements

The author is much indebted to Mr G.L.G.A. Kortekaas, town archaeologist of Groningen, Mr J. Schoneveld, B.A.I., Mr K. Helfrich of the Stichting Monument en Materiaal (M & M), who supplied much information about the excavations, and Mr P.H. Broekhuizen (M & M) who took the photographs. The drawings were made by Mr H.R. Roelink and Mr J. Smit. The English text was corrected by Ms A.C. Bardet and Mrs G. Entjes-Nieborg assisted in preparing the manuscript.

References

Casparie, W.A. (1988). Het houtgebruik. In *Kattendiep deurgraven. Historisch-archeologisch onderzoek aan de noordzijde van het Gedempte Kattendiep te Groningen* (ed. P.H. Broekhuizen, A. Carmiggelt, H. van Gangelen & G.L.G.A. Kortekaas), pp. 132-140. Groningen: Monument en Materiaal.

Casparie, W.A. (1991). Houtgebruik in het vroeg-middeleeuwse grafritueel in Noord-Nederland. *Paleo-Aktueel* **2**, 103-107.

Casparie, W.A. & Helfrich, K. (1993). De houten voorwerpen. In *Van boerenerf tot bibliotheek. Historisch, bouwhistorisch en archeologisch onderzoek van het voormalige Wolters-Noordhoff Complex te Groningen* (ed. P.H. Broekhuizen, H. van Gangelen, K. Helfrich, G.L.G.A. Kortekaas, R.H. Alma & H.T. Waterbolk), pp. 431-473. Groningen: Monument en Materiaal).

Lanting, J.N. (1990). De ouderdom van de houten gebouwen onder de St.-Walburg- en Martinikerk. In *Groningen 1040. Archeologie en oudste geschiedenis van de stad Groningen* (ed. J.W. Boersma, J.F.J. van den Broek & G.J.D. Offerman), pp. 155-174. Bedum: Profiel.

Neugebauer, W. (1975). Arbeiten der Böttcher und Drechsler aus der mittelalterlichen Bodenfunden der Hansestadt Lübeck. *Rotterdam papers II. A Contribution to Mediaeval Archaeology*, 117-137.

Trees, Woodland and Woodmanship in Early Mediaeval Ireland

A. O'SULLIVAN

The Discovery Programme, 13-15 Hatch Street Lower, Dublin 2, Ireland

Summary

The evidence for woodmanship in mediaeval Ireland is discussed. Palynological evidence dispels the continuing myth that there were vast mediaeval woods. Nonetheless, place name and documentary sources indicate discrete areas of woodland. At these places underwood was probably managed, livestock sheltered and fattened, berries and nuts gathered and timber felled. A range of historical texts and archaeological excavations support these claims. Woodland was an important aspect of the agricultural landscape and economy.

Introduction

The role of woodland resources in early mediaeval Ireland (6th-12th century AD) can best be reconstructed through a combination of historical, archaeological and palynological evidence. In terms of assessing the actual character and extent of woodland, of primary importance is the analysis of pollen cores taken from raised bogs and lake muds. In summary, the landscape would have been generally open, with only scattered oakwoods (*Quercus* spp.). Ash (*Fraxinus excelsior* L.) and hazel (*Corylus avellana* L.) would have been common in hedgerows and secondary woodland. Willows (*Salix* spp.) and alders (*Alnus glutinosa* L.) would have been present on wetland margins, while other trees such as yew (*Taxus baccata* L.), holly (*Ilex aquifolium* L.), elder (*Sambucus nigra* L.), hawthorn (*Crataegus monogna* Jacq.), blackthorn (*Prunus spinosa* L.), birch (*Betula* spp.) and rowan (*Sorbus aucuparia* L.) would also have been widespread (Mitchell, 1986). Scots pine (*Pinus sylvestris* L.) may have survived in isolated pockets throughout the period, while elm (*Ulmus glabra* Huds.) seems to have nearly disappeared as a timber tree. Such species as beech (*Fagus sylvatica* L.), lime (*Tilia* spp.), hornbeam (*Carpinus betulus* L.) and sycamore (*Acer pseudoplatanus* L.) were absent from mediaeval Ireland.

Cartographic and place name evidence

In complete contrast, place name and literary evidence would tend to suggest a picture of more extensive woodland cover. The Old-Irish word *doire*, denoting oakwood, is a very common element in Irish place names. Place names with *doire, eo, beith* all signify the presence of either single trees or woods of oak, yew and birch respectively. Idiomatic historical phrases can also be intriguing. One mediaeval text describes cattle being hidden from raiding armies in 'dense woods of lofty foliage' in Slieve Aughty, a region in County Clare which has profuse numbers of *doire* townland names. Perhaps some regions may have retained a largely woodland character into the late mediaeval period. A Tudor map completed in *c.*1563 shows the midland counties of Laois and Offaly and depicts named woodland in river valleys, on isolated bog islands and lower mountain slopes (Smyth, 1982). However cartographic research by Rackham (1986) on the more reliable Civil Survey maps indicates that by 1654 the percentage of woodland cover had fallen in some counties as low as 2.1%.

Woodland in the landscape

Whatever their extent, the exploitation of early mediaeval woodlands can be traced in the historical annals, saint's lives and early Irish law-texts that were being compiled by contemporary monastic scholars in the eighth and ninth centuries AD. These record, as incidental detail, an immense wealth of social and economic history. It is clear from these texts that woodland could be owned, exchanged and managed in a variety of ways and was treated as a distinct, valuable zone within a highly structured agricultural landscape (O'Sullivan, 1992). 'Legal standard woodland' or enclosed woodland was one of the natural features which could increase the value of agricultural land (Mac Niocaill, 1971). The type of fence used in enclosing such woodland was specifically described in the late seventh century *Bretha Comaithchesa* as the *dairime* or oak fence. This seems to have been made by partially felling young trees and bending them over to create a laid hedge. Such a dense construction would be more efficient at keeping out browsing animals and may be indirect evidence for the practice of woodland management (O'Corrain, 1983). In common with other types of land, woodland was typically owned by the family kin-group but all law-abiding members of the *tuath* had certain rights within the wood. These included rights to the collection of firewood or nuts and the cutting of a certain amount of rods for wattling.

An insight into the relative values of different species can also be seen in the seventh century tree list preserved in the *Bretha Comaithchesa*. This lists 28 different native species and classifies them into four groups (Kelly, 1976). Significantly these classifications are based on economic rather than botanical properties. The first grade, the *airig fedo* or the 'nobles of the wood' includes oak because of its acorns, its size and appearance, hazel for its nuts and rods, holly possibly for the use of its leaves for winter fodder, yew for the production of fine or 'noble' wooden artifacts and apple for its fruit and bark. There is also evidence for the symbolic importance of some individual trees, especially oak and yew (Lucas, 1963). These sacred trees or *bile* could be associated with tribal groupings at royal inauguration sites. Their destruction during the frequent cattle raids of the time by neighbouring rivals was seen as a gross affront. Named sacred trees could also be associated with individual monastic centres or saints. Thus the value of trees was not always an economic one.

Pannage and wood-pasture

It is also likely that animals were maintained in selected areas of woodland. The mediaeval Irish annals typically provide a dry, laconic account of various battles, deaths and ecclesiastical or political successions. However, references are also included to climatic or other natural phenomena. Between AD 672 and 1155 there are as many as 23 separate annalistic references to prodigious annual harvests of acorns (mast). For example, in AD 836 the crop of mast was so heavy it blocked streams, in AD 969 eight bags were collected from beneath each tree. In AD 985 the crop was so great it lasted until the following year, while in AD 1038 the annals record that the crop was so abundant that even the runts of the swine were fattened. It has traditionally been believed by historians that these crops of mast were primarily used for fattening pigs, although the unpredictability of the harvests must have occasionally presented problems. Certainly pig-meat in the form of fresh pork and salted bacon was very popular amongst the mediaeval Irish. Cattle were also occasionally kept in woodland for winter grazing or for protecting them from cattle raiding expeditions.

Fruit and nuts

A reference in the *Annals of the Four Masters* for the year AD 1031 gives a clue to another likely use of the yearly harvests of woodland fruit and nuts. It reads as follows *'a measure of oaten grain, or a third of black-red sloes, or of the acorns of the brown oak, or of the nuts of the fair hazel hedge, was got without stiff bargaining, at Armagh for one penny'* (O'Donovan, 1856). It may be that such berries and nuts were being used as a cash crop to be sold at town markets. Certainly apples were particularly highly valued, monasteries had their own orchards and secular tenants who moved from their property could be compensated by law for the apple trees that they had sown. Otherwise, despite the perception of the value of woodlands, there seems to be little evidence for the deliberate plantation of trees. After the establishment of towns proper in the tenth century, the bulk of trade in most woodland materials must have been with the major centres such as Dublin, Cork, Waterford and Limerick. Excavations in mediaeval Dublin have discovered evidence for fruit processing pits, while the species of mosses that were used for personal hygiene in latrines could only have been specially gathered from the shaded tree-trunks, rocks and soils of deciduous woodlands (Dickson & Dickson, 1987).

Underwood

Otherwise the most important use of woodland must have involved the production of long, straight hazel, ash and willow rods. House walls, property fences and wooden pathways would all have been made of woven hurdle panels and would have required vast amounts of quality rods or underwood. Archaeological evidence for these structures is illustrated particularly in the settlement of Viking and mediaeval Dublin. The amounts of hurdle panelling used should indicate a close relationship with the nearby, native hinterland, possibly involving organized trading of woodland products. Underwood could have been valued in areas of land, related to the labour required to crop it. As late as AD 1340 the Holy Trinity Priory in Dublin paid two men a shilling a day to crop underwood in the woods of Clonken only five miles south of Dublin (O'Neill, 1987).

Deer Park Farms, County Antrim

The most important recent archaeological research on this subject has been carried out by Marie Neill on the wood found in an excavation at Deer Park Farms, Co. Antrim. The site was an early mediaeval ringfort or settlement, at the base of which waterlogging had preserved a number of circular wooden houses dated to the seventh century AD. These had double walls of post-and-wattle in which a highly complex basketry-like weave was used. This technique would have required about 6,000 rods for the largest house, with varying amounts for the other structures. The upright posts were simply chosen for size from mixed woodland with alder, birch, ash, oak and holly all being used. In contrast the horizontal rods were entirely of hazel and tree-ring analysis has shown that these were possibly being taken from a managed area of woodland at eight year intervals (M.E. Neill, pers. comm.).

Fergus estuary fishweir, County Clare

Intertidal surveys on the Fergus estuary by the author have led to the recent discovery on the mudflats of a wooden fishweir dated to the sixth century AD (O'Sullivan, 1993). This structure was a post-and-wattle fence designed to guide fish swimming down with the tide into a trap. The predominant species used for the posts and rods in this structure were willow, alder and hazel (Fig. 1). The willow and alder would most likely have grown on the fringes of the surrounding saltmarshes, and provide a useful indicator of the local landscape. Preliminary analysis of this wood suggests that immature willow and hazel branches were simply being selected for size from adventitious growth. Given the restricted ages in the case of the alder posts and rods, it is perhaps possible these were taken from a coppiced alder stool, although the small size of the sample makes this a tenuous assertion.

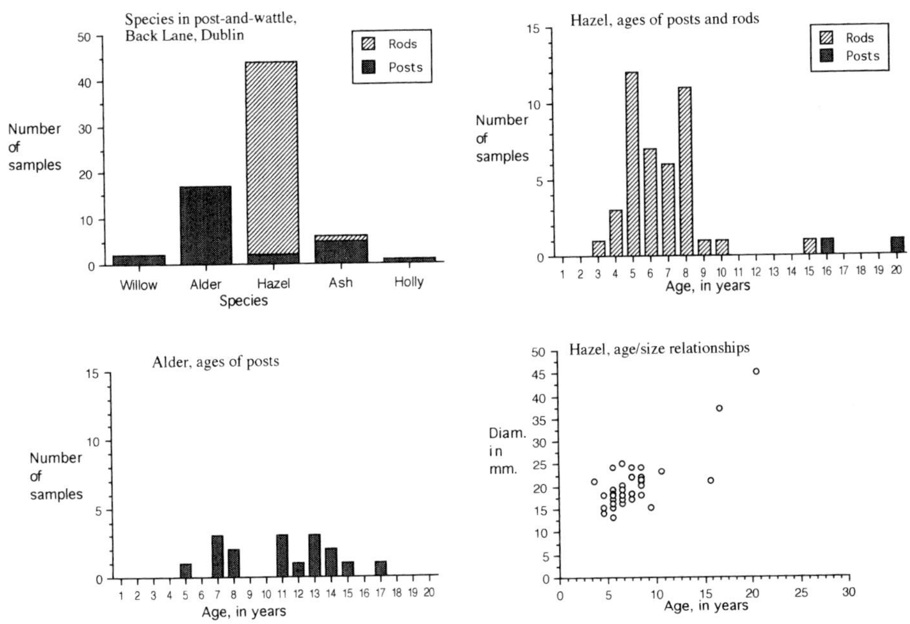

Fig. 1. Wood analysis on sixth century fishweir, Fergus estuary, County Clare

Back Lane, Dublin

Recent excavations at Back Lane, in Dublin by Claire Walsh uncovered amongst other structures, a rectangular house dated to the early thirteenth century. Close attention was paid during the excavation by the author to the recording of the surviving post-and-wattle walls and preliminary analysis of the wood samples has been completed. The house can be classed as a Dublin Type 1 house, with internal roof supports and tripartite dividing walls. This is the classic house type from mediaeval Dublin, being built from the earliest Viking origins up to the Anglo-Norman period (C. Walsh, pers. comm.).

Clearly a range of species were being used for the upright posts, including willow, alder, hazel, ash and holly (Fig. 2). When the tree-ring or age patterns of these various species were examined, no concentrations were apparent. Seemingly these posts were again simply being taken from suitably sized poles of any species from mixed woodland, although alder does predominate. In contrast, a clear preference for hazel was exhibited in the rods sample. The distribution of ages in these hazel rods is more difficult to interpret, peaks can be seen at five and eight years, with over 80% of the sample falling between these ages, while the age/size distribution suggests a tight cluster. Hazel underwood of a certain size was being cropped between eight and five years growth, but not necessarily in the same area of woodland. It may be that this represents the choice of rods for size rather than a formal cycle of woodland management.

There would seem to be restricted archaeological evidence yet for the formal woodland management procedures known from late mediaeval England (Rackham, 1982; Crone, 1987). However the huge amounts of underwood that would have been necessary for constructing and renewing the houses in early Dublin, would surely have required some form of woodland management, perhaps of a more complex nature not readily identifiable in comparison to more recent practices (O.

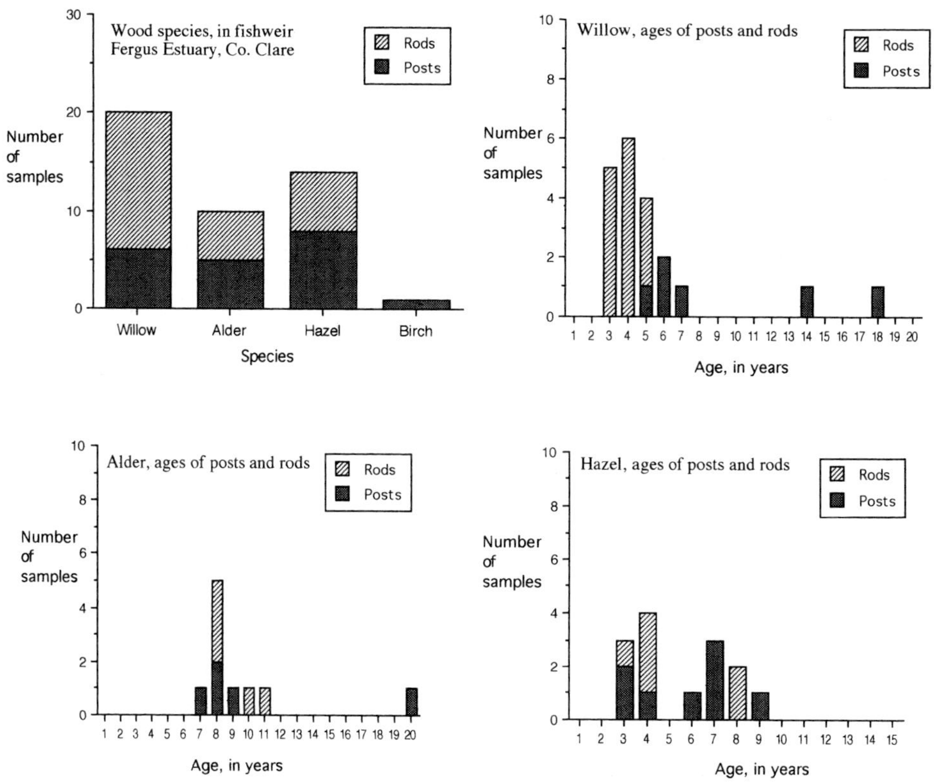

Fig. 2. Wood analysis on thirteenth century house, Back Lane, Dublin.

Timber

The second main use of woodland involved the growth, felling and conversion of large trees into timbers for making houses, trackways, waterfronts, watermills and a variety of other structures. Oak was easily the most commonly used species in early mediaeval Ireland, the strength and durability of its wood made it especially suitable for loadbearing and architectural uses. It is more difficult to say if standard oaks were being grown within a managed system. Certainly the unique value of oak was recognized in the law-texts. Damages up to the value of a milch cow were due to the owner of a tree if its branches were removed or if it was felled without permission. Fines of a cowhide were imposed on anybody illegally removing oak bark for tanning leather. Indeed the eighth century wisdom text *Uraiccecht Bec*, describes the status of the '*accurate wright of oaken houses*' as being equivalent to the lowest grade of nobility (Binchy, 1958).

Woodmen

The felling and transport of timber from source to the mediaeval towns could have involved specialized woodmen. The process of felling large trees by specialized craftsmen for use in timber seems to be described in one passage in Cogitosus's seventh century *Vita Prima Sanctae Brigitae*. In the account, a large, tall tree was felled *'by those who used to ply their trade in a timber forest'* and thereafter specialized equipment or *'skilled devices'* were used to drag it out of the wood to the *'appointed place'* (Connolly and Picard, 1987). In the undated *Life of St Samthan*, carpenters are stated to have transported pine poles long distances home in carts drawn by oxen. Woodmen would have been the first link in chain of timber trade, involving vehicles and draught animals to transport roughly prepared planks and beams to a construction site. In later periods the trade of timber was managed by the native Irish. Indeed the Civil Regulations of Waterford implied that the '*wode bote men*' were troublesome, so much so that by AD 1458 an oath of loyalty was required from '*bene maistres of wod botes and botemen*' (O'Neill, 1987). The conveyance of actual timbers across large distances could also have been accomplished by boats. Adomnán's seventh century *Life of Columba* (Anderson & Anderson, 1961) describes how pine was imported onto the island of Iona from mainland Scotland in the following terms, '*when dressed timbers of pine and oak for a long ship were being drawn overland, and timbers were conveyed from the great house, as well as for ships... with boats and currachs to tow the timber through the sea*'.

Anglo-Norman carpentry at Woodquay

The best evidence for the use of timber in early mediaeval Dublin is the wooden river waterfronts at Woodquay, the earliest of which has been dated to AD 1210. By this time native Irish carpentry styles were beginning to be used in an urban context, with the introduction of mortise and tenon joints, through-splayed scarfs and heavy hewn timbers. These structures have been well described by Dr Pat Wallace and can be placed firmly within the north European tradition of front braced waterfronts (Wallace, 1982). Their construction involved a series of braced oak uprights in a baseplate supporting a wall of planks which held back the estuarine clays, gravels and urban deposits dumped behind.

Excavations carried out recently on Winetavern Street by Andy Halpin have enabled an examination of the timbers used from the point of view of woodmanship. Particular attention during the recording process was paid to species identifications, ages, growth rates, position of sapwood, knots and general appearance of the wood. With this information it should now be possible to reconstruct tentatively the types of trees used and the appearance of the original woodland. Baseplates were of boxed-heart section and seem to have been taken from quite straight-grained oak trunks, 40cm in diameter which were free of knots or branches for most of their length. The sapwood or waney edge has been left on the corners and they were taken from trees under 100 years of age. Smaller lengths of knotty trunks, braced in position by either radially cleft oak beams or actual branches, held back finely split oak planks. The uprights were most likely taken from mature coppiced poles about 50 years old or the topwood from larger trees.

Conclusion

In conclusion, it can be said that woodland resources, although prosaic, were a vital ingredient in the early mediaeval Irish economy. Woodland was clearly perceived by contemporary scribes as a valuable distinct zone, occasionally peripheral to other concerns, but always of importance in the agricultural landscape. There is a range of evidence for the management of underwood in some form, for the foddering of livestock on the woodland floor and for the organized gathering of fruit, nuts and mosses for exchange in the growing urban markets. Increasing scarcity of timber quality oaks seems even to have been reflected in legal restriction. However it is also clear that no single discipline, archaeology, early Irish history, palynology or dendrochronology is capable of elucidating these elements on its own. Through an holistic combination, the subject of trees, woodland and woodmanship in early mediaeval Ireland can be more usefully investigated.

References

Anderson, A. & Anderson, (1961). *Adomnán's Life of Columba.* Edinburgh.

Binchy, D.A. (1958). The date and provenance of Uraicecht Bec. *Eriú* **18**, 44-54.

Crone, A. (1987). Tree-ring studies and the reconstruction of woodland management practices in antiquity. *Proceedings of the International Symposium on Ecological Aspects of Tree-ring Analysis* (ed. G.C. Jacoby & J.W. Holmbeck), 327-336.

Connolly, J. & Picard, J.M. (1987). Cogitosus: life of Saint Brigit. *Journal of the Royal Society of Antiquaries of Ireland* **117**, 5-27.

Dickson, C.A. & Dickson, J.H. (1987). The plant remains. In *Archaeology and Environment in Early Dublin* (ed. G.F. Mitchell), pp. 22-32 Dublin: Royal Irish Academy.

Kelly, F. (1976). The Old Irish tree-list. *Celtica* **11**, 107-124.

Lucas, A.T. (1963). The sacred trees of Ireland. *Journal of the Cork Historical and Archaeological Society* **68**, 16-54.

Mac Niocaill, G. (1971). *Tír Cumaile. Eriú 22, 81-86.*

Mitchell, F. (1986). *The Shell Guide to Reading the Irish Landscape.* Dublin: Country House.

O'Corrain, D. (1983). Some legal references to fences and fencing in early historic Ireland. In *Landscape Archaeology in Ireland* (ed. T.E. Reeves-Smyth & F. Hamond). Oxford: British Archaeological Reports, British Series 116.

O'Donovan, J. (1856). Annála Ríoghachta Eireann. *Annals of the Kingdom of Ireland, by the Four Masters* 7 Vols. Dublin.

O'Neill, T. (1987). *Merchants and Mariners in Mediaeval Ireland.* Dublin.

O'Sullivan, A. (1992). Trees and woodland in early mediaeval Ireland: an ethnohistorical approach. *News WARP* **11**, 3-7.

O'Sullivan, A. (1993). Intertidal survey on the Fergus estuary and the Shannon estuary. *Discovery Programme Reports 1*, pp. 61-68. Dublin: Royal Irish Academy/Discovery Programme.

Rackham, O. (1982). The growing and transport of timber and underwood. In *Woodworking Techniques Before AD 1500* (ed. S. McGrail), pp. 199-218, Oxford: British Archaeological Reports, International Series 129.

Rackham, O. (1986). *The History of the Countryside.* London: J.M. Dent.

Smyth, A.P. (1982). *Celtic Leinster.* Dublin: Irish Academic Press.

Wallace, P.F. (1982). Carpentry in Ireland AD 900-1300 - the Woodquay evidence. In *Woodworking Techniques Before AD 1500* (ed. S.McGrail), pp. 263-299. Oxford: British Archaeological Reports, International Series 129.

Pollards in Art

C-A. HÆGGSTRÖM

Department of Botany, P.O. Box 4, FIN-00014 University of Helsinki, Finland

Summary

The paper is a brief discussion of the representation of pollards in European art from mediaeval times to the 20th century.

Leaves of deciduous trees have been used as winter fodder for domestic animals since the Neolithic Age (Rasmussen, 1990; 1991). Loppings were often gathered through pollarding and pollards could be useful also for firewood, fences, faggots and for carpentry (Hæggström, 1992).

Since written evidence on pollarding from earlier times is scarce, or even lacking, some knowledge can be gathered from works of art. With a few exceptions, *e.g.* Glück (1951), Makowski & Buderath (1983), Rud (1988) and Ellenius (1992), the literature on art does not, however, deal with the pollards depicted in works of art. Only a few botanical papers deal with such pollards; examples are Andersen (1985) and Slotte (1992).

The oldest works of art with pollards I have hitherto seen are frescos and reliefs on gravestones and altars from the first centuries AD. Mediaeval church art abounded in conventionalized pollards.

The earliest true-to-life landscape paintings with pollards are in *Très Riches Heures of Jean Duc de Berry,* painted by the Limbourg brothers in 1411-16 (Fig. 1) (see Longnon & Cazelles, 1969). Many Dutch, Flemish and German artists of the 15th to 18th centuries often painted pollards (Fig. 2). Of the 19th century English artists, William Turner and John Constable (Fig. 3), as well as some of the French impressionist (Fig. 3) can be mentioned. Works of art of Nordic artists depicting pollards are illustrated in Fig. 5-6.

While church art mostly depicts imaginative landscapes and pollards, the landscape painters, beginning with the Limbourg brothers, depict life-like landscapes with real pollards. The use of cut branches for fences, faggots and firewood can be seen in several works of art. In some cases the tree species can be determined, and in a few cases the whole cultural landscape with its pollards can be traced in paintings. A full list of paintings illustrating pollards may be obtained from the author.

Acknowledgements

Permission to reproduce the works of art illustrating this paper from Musée Condé, Chantilly (photograph by Giraudon); Kunthistorisches Museum, Vienna; Yale Center for British Art, New Haven Connecticut; Musée du Louvre, Paris; Ny Carlsberg Glyptotek, Copenhagen; and Turku Art Museum, Finland (photograph by Gustaf Welin) is gratefully acknowledged.

References

Andersen, S.T. (1985). Natur- og kulturlandskaper i Næsbyholm Storskov siden istiden. (Natural and cultural landscapes in Næsbyholm Storskov since the Ice Age.) *Fortidsminder* 1985, 85-107.

Fig. 1. The Limbourg brothers: June in the 'Très Riches Heures du duc de Berry', 1411-1416. Row of pollards along the Seine with Palais de la Cité of Paris in the background. Musée Condé, Chantilly. Photograph: Giraudon.

Ellenius, A. (1992). *Landskappsbilden. Ur den europeiska naturkänslans historia,* 86 pp. Stockholm: Natur och Kultur.

Glück, G. (1951). *Das Grosse Bruegel-Werk,* 124 pp. + plates. Wien: Anton Schroll & Co.

Hæggström C-A. (1992). Wooded meadows and the use of deciduous trees for fodder, fuel, carpentry and building purposes. *Protoindustries et histoire des forêts. GDR ISARD-CNRS. Les Cahiers de l'ISARD* **3,** 151-162. Toulouse.

Longnon, J. & Cazelles, R. (1969). *The Très Riches Heures of Jean, Duke of Berry. Musée Condé, Chantilly.* 224 pp. George Braziller Inc. / Secaucus, NJ: The Wellstreet Press.

Makowski, H. & Buderath, B. (1983). *Die Natur dem Menschen untertan. Ökologie im Spiegel der Landschaftsmalerei.* 309 pp. München: Kindler Verlag.

Rasmussen, P. (1990). Leaf foddering in the earliest Neolithic agriculture. Evidence from Switzerland and Denmark. *Acta Archaeologica* **60,** 71-86.

Rasmussen, P. (1991). Leaf-foddering of livestock in the Neolithic: Archaeobotanical evidence from Weier, Switzerland. *Journal of Danish Archaeology* **8,** 51-71.

Rud, M. (1988). *The Bayeux Tapestry and the Battle of Hastings 1066,* 103 pp. Copenhagen: Christian Eilers.

Slotte, H. (1992). Muralmålning visar lövtäkt på 1400-talet i Uppland. *Svensk Botanisk Tidskrift* **86**, 271-273.

Thorsen, S. (1988). Pilehegn og risgærder, 15 pp. *Lolland-Falsters Historiske Samfunds Årbog 1988.*

Fig. 2. Pieter Bruegel the elder (*c.* 1525-1569): 'Düsterer Tag' (Gloomy Day), 1565. Panel with pollarding. Kunsthistorisches Museum, Vienna.

Fig. 3. John Constable, R.A. (1776-1837): 'Parham Mill, Gillingham', *c.* 1826. Willow Pollards. Yale Center for British Art, New Haven Connecticut, Paul Mellon Collection.

Fig. 4. Jean Baptiste Camille Corot (1796-1875): 'The bridge at Mantes, France', 1868-70. Freshly repollarded willow. Musée du Louvre, Paris.

Fig. 5. Peter Christian Skovgaard (1817-1875): 'Cart-road with willows at Vognserup', 1847. Willow (or poplar?) pollards. Ny Carlsberg Glyptotek, Copenhagen.

Fig. 6. Victor Westerholm (1860-1919): 'Winter landscape', 1897. Oil on canvas 100 x 58cm. Turku Art Museum, Finland. Photograph:Gustaf Welin.

Plants and People in Oral Tradition: Folklore in the Scottish Highlands

MARGARET BENNETT

School of Scottish Studies, 27 George Square, The University of Edinburgh, Edinburgh, EH8 9LD

Summary

An account is presented of how plants have been viewed in folklore with particular reference to aspects of Gaelic tradition.

This paper is written not from the point of view of the botanist but from that of the folklorist, whose study of plants focuses upon traditional knowledge, beliefs and legends about plants, and also people's attitudes to them. From its very infancy, the science of botany has made use of a large body of information that was handed down from generation to generation in various cultures throughout the world. Much has been shared with other disciplines, such as medicine, which documents many well-known 'discoveries' related to plantlore. For example, the ancient practice of chewing willow bark to relieve 'agues' and fevers led to the identification of the active ingredient we now know as salicylic acid, or aspirin. In the closing decade of the twentieth century, it is reassuring to know that botanists remain open to considering folk-knowledge of plants, in an age when their science has attained the sophisticated finesse of the micro-chip. In the study of folklore, 'plants and people' have had a long-standing meeting-place.

From within our own culture we may take for granted aspects of traditional knowledge. On a practical level, we learn at an early age that docken (*Rumex*) leaves can counter nettle (*Urtica*) stings, or of the diuretic properties of dandelions (*Taraxacum*). Yet when we step outside of our own culture and into one less known to us, we may raise an eyebrow in surprise, or smile at the apparent naivety of the beliefs relating to trees, plants and nature in general. The idea that spirits inhabit trees is sometimes associated with so-called 'primitive cultures'. Thus, a belief that 'if you cut down a tree you will offend the spirit that inhabits it' is not one you are likely to hear in Scotland. Nevertheless our present ideas of conservation are not so far removed from those of other cultures or ages: in ancient Hebrew law it was an offence to hew down trees, and in some areas of Germany the undue cutting of trees that marked village boundaries was strictly forbidden.

Less familiar ideas about plants, such as the Malayan belief that the coconut has eyes, and so never falls on anyone's head, or the Indian idea that you can marry a tree, may seem totally alien to our Scottish culture, but when we consider some of our own attitudes it becomes clear that ours are no more logical. Consider, for example, the many streets in Edinburgh that are punctuated by rowan trees. As a Highlander, I found that surprising because I have long associated rowans with the Highlands. Nevertheless, the belief that the rowan, or mountain ash (*Sorbus aucuparia* L.) is lucky, or averts evil, is widespread throughout Britain, with actual explanations varying according to geographic location. It is, perhaps, paradoxical that the strongest belief in the supernatural properties of the rowan exists in the

Highlands and Islands of Scotland, where the strictest form of Presbyterianism is also to be found. Despite the fact that the staunchly Calvinistic Reformed Church succeeded in wiping out many of the so-called un-Christian customs and beliefs, this is one area in which the church has made little or no impression. To this day, it is quite common to find devout churchgoers with rowan trees near their houses which will not, under any circumstances, be violated. Without hesitation they will state that their forebears originally planted rowans to keep away witches (based on field work interviews on Skye in 1990).

When I moved to Edinburgh in 1983, I could not say that I planted my little rowan sapling in the front garden exactly for those reasons. I preferred to acknowledge that the planting was more a link with generations of tradition. In May 1990, however, when I returned from a weekend's absence to find it cut down by my new neighbour from London, I was forced to confront the real significance of the tree to my family. Why should there be feelings of distress to find it chopped down? Was it simply the loss of a much-loved tree? She had, after all, removed the dandelions, pulled out the marigolds and poured weed-killer on my anemones that strayed onto the path. Clearly the rowan evoked the strongest of all feelings, and left me with the question whether it was the violation of a symbol that was my daily connection to my Gaelic cultural background, or was it something more? The psyche of the individual has little to do with logic, and even less to do with education.

My personal interest in plantlore is lifelong and may be attributed to my upbringing in a Gaelic society where age-old traditions encompassed daily practices that testified to ancient wisdom. This was not textbook learning but the tested and tried ways of generations of Gaelic tradition-bearers from my own family and community. Aside from the daily 'sayings' about plants, however, we owe much of our knowledge to early travellers who observed regional plantlore and recorded it in writing. One of Scotland's most significant travel writers was Martin Martin, a native of the Isle of Skye, whose *Description of the Western Isles circa 1695* (Martin, 1716) encompasses an enormous legacy of folklore from these Scottish islands. The fact that Martin is not better known has more than a little to do with his better known successor, Dr Samuel Johnson, who acknowledges Martin's *Description* as his inspiration, then promptly pronounces it a badly written book, thus diverting generations of readers to Johnson's own work. That work is better known than any other, despite the fact that his methods of research were entirely questionable. Crucially, he did not speak the language of the people he travelled among, and when they made little sense of his questions his skilful journalism belittled their peasant ways.

Martin Martin's matter-of-fact manner of noting the daily practices he observed on his travels in the Hebrides sheds permanent light upon the significance of plants in the lives of the Scottish Gaels. While he does not categorize or classify either the plants or their uses, what emerges is a body of information affirming beliefs and practices that were extant in Martin's time, and which, to a large extent, can still be recorded from oral tradition in Gaelic Scotland today. (The School of Scottish Studies of Edinburgh University has several hundred recordings of the information about plantlore from all over Scotland. See also Cameron (1883)). His work reflects the belief firmly held by Gaels that, wherever an ailment occurred, there nature provided a plant that would offer treatment. Some of the practices noted by Martin can also be compared to similar ones that date back to ancient Greece, as some of the early Gaelic physicians, notably the Beatons, had studied the work of Hippocrates and had manuscripts in which parts of it were translated into Gaelic (Bannerman, 1986).

In his introduction, Martin Martin (1716) stated:

'The wise Solomon did not think it beneath him to write of the meanest plant, as well as of the tallest cedar. Hippocrates was at the pains and charge to travel foreign countries, with a design to learn the virtues of plants, roots, etc. I have in my little travels endeavoured among other things, in some measure to imitate so great a pattern; and if I have been so happy as to oblige the republic of learning with anything that is useful I have my design. I hold it enough for me to furnish my observations without accounting for the reason and way that those simples produce them. This I leave to the learned in that faculty; and if they would oblige the world with such theorems from these and the like experiments, as might serve for rules upon occasions of this nature, it would be of great advantage to the public.'

When the World Health Organization launched their 10-year study into herbal remedies in 1983, it seemed to me that the work of Martin Martin, completed almost three centuries before, and indeed the works of other Gaelic writers, should be more closely studied. To cite only a few examples from the range Martin recorded, he noted in Skye 'the tops of nettles, chopped small, and mixed with a few whites of raw eggs, applied to the forehead and temples by way of a frontel, is used to procure sleep'. In Lewis he recorded a home-remedy which involved the roots of nettles being boiled in water along with the roots of reeds, then fermented by adding yeast 'which they find beneficial for the cough.' Besides cures for ailments as diverse as migraine, fever, worms and hearing loss, he noted tonics such as herbs boiled in lamb broth for consumption, and preventatives such as the root of the bitter vetch (*Lathyrus montanus* (L.) Bernh.), which was said to stave off drunkenness.

Magical cures are noted alongside more practical ones, such as the Moluka bean, used as an amulet against witchcraft or evil eye (Martin, 1716; Nelson, 1983). The white bean was especially valued, as it changed colour to black when the wearer was in danger. Some plants even had the power to prevent second sight, such as the elegant St. John's wort, (*Hypericum pulchrum* L.) which Martin observed on the Isle of Harris, worn in the armpit of a man who was said to 'suffer' from second sight. Sure enough, he never again experienced it, and thus the belief in the power of the plant was perpetuated.

Another writer who recorded Gaelic oral tradition was Alexander Carmichael, a late nineteenth century collector of sayings, rhymes, charms, and legends, many of them about plants, spanning a 'folk memory' of over 150 years (Carmichael, 1900-1971). Among his collection are religious legends about plants, such as the one concerning lint (flax; *Linum usitatissimum* L.): the hands and feet of Christ were said to have been bound with lint flax, when He was taken down from the cross, and before He was carried to the grave. Ever after, lint has been referred to in Gaelic as *'lìon beannaichte'*, the blessed lint, or *'lìon naomh'*, the sacred lint, and there is a saying that a person who would steal lint would be guilty of as heinous a sin as anyone sinning against the Holy Ghost.

'Theft of lint or theft of seed,
Two thefts from which no peace nor relief can be,
Till the broad world comes to an end
The thief of lint shall get no respite.'

Some plants are particularly rich in traditional lore, such as the bramble (*Rubus fruticosus* L.). It was highly valued as a fruit for food, as medicinal tea made from the leaves, as a dye made from the berries and/or the root, as an amulet or good luck token and it was sometimes put above byre lintels to ward off evil. It was even the subject of a Christian legend, in Gaelic *'An druise beannaichte'*, the blessed bramble, in which it is said that a branch of it was the wand with which Christ hastened the ass when going to Jerusalem. It was also said to have been the rod with which He drove the money-changers out of the temple. More locally, it is the subject of a Gaelic proverb:

> 'He who would go in the bramble for me,
> I would go in the thorn for him.'

It is not only the inhabitants of rural parts of the Highlands who hold beliefs connected with plants, however, as these sayings are widespread, indicating the importance of plants in the lives of ordinary people. To continue the example of the bramble, it is regarded in many areas of Britain as having healing powers, even magical properties, such as the ability to cure rickets if the sufferer is taken between the arches (Radford and Radford, 1969). It is also efficacious in curing skin troubles and whooping-cough, and in Cornwall the bramble leaves were used in an elaborate ritual to heal burns, scalds and inflammations: nine leaves were floated in a basin of water drawn from a holy well or pure spring. Then, one by one, the leaves were passed over and away from the affected part while this was repeated three times:

> 'There came three angels out of the east,
> One brought fire and two brought frost.
> Out fire, and in frost,
> In the name of the Father, Son and Holy Ghost.'

Religious legends also refer to the bramble as one of the plants from which the Crown of Thorns was made. In some areas children are discouraged from picking brambles or blackberries after Michaelmas Day (September 29, or Old Style, October 11) as Satan is reputed to have cursed the fruit when he was cast out of Heaven. According to legend, he fell into a bramble bush, so on every anniversary after that, it is said that the Devil spat on them and that they bring only bad luck or even death.

Although this paper can only touch on aspects of it, the subject of traditional plantlore is enormous, as can be gathered from the many collections from both printed and oral sources (Hilger and Mondloch, 1972; Bennett, 1991; Bennett, 1992). It is not, however, a subject which can be relegated to the past, as there are still a great many traditions attached to plants even today. (Bennet, 1992). In recent years I have offered the subject for discussion to students at the University of Edinburgh, and I am left in no doubt that beliefs and attitudes to plants are very much part of our present way of life. In 1992, when discussing the customs and beliefs surrounding death, a young student wrote in her essay that her grandmother would never have snowdrops in the house 'because they brought death with them.' Other students insisted that they would never have red and white flowers together, as they foretold death. Even in our modern hospitals, nurses will make every effort to avoid putting those two colours in the one vase, for despite up-to-date medical training and textbook expertise, there is a much deeper level of learning that comes into force when dealing with such matters.

No matter how scientifically we can disprove many of the above examples of traditional lore, our proof is of little importance to those who perpetuate the sayings, beliefs and practices as part of their way of life. As with a great many customs, they add colour and meaning to aspects of life that might otherwise become absorbed in the humdrum, everyday routine that already occupies too much of our human psyche. If, perchance, we keep a four-leafed clover for luck, or plant a houseleek to safeguard against fire, or, in a more frivolous moment, we want to find out if a friend likes butter by placing a buttercup under her chin, or to tell the time by blowing the seeds of a dandelion, or to turn our thoughts to whether 'he loves me/ he loves me not' by plucking the petals from a daisy, then perhaps we are fulfilling some other need, possibly one that is at the centre of an intangible, inexplicable relationship between plants and people.

References

Bannerman, J. (1986). The Beatons: A Medical Kindred in the Classical Gaelic Tradition, pp. 89-119. Edinburgh: John Donald.

Bennett, M. (1991). Plant lore in Gaelic Scotland. In *Flora of the Outer Hebrides* (ed. R.J.Pankhurst & J.M.Mullin), pp. 56-60. London: Natural History Museum Publications.

Bennett, M. (1992). *Scottish Customs from the Cradle to the Grave,* 298 pp. Edinburgh: Polygon.

Cameron, J. (1883). *Gaelic Names of Plants,* 130 pp. Edinburgh: William Blackwood and Sons.

Carmichael, A. (1900-1971). *Carmina Gadelica: Ortha nan Gaidheal,* Volumes **1** (1900) and **II** (1900), Edinburgh: T. & A. Constable. Volumes **III** (1940) and **IV** (1941) [ed. J. Carmichael Watson], Edinburgh: Oliver and Boyd. Volumes **V** (1954) and **VI** (1971) [ed. A Matheson], Edinburgh: Scottish Academic Press.

Hilger, M.I. & Mondloch, M. (1972). A source list in ethnobotany. *New York Folklore Quarterly* **28**(1), 61-78.

Martin, M. (1716). *Description of the Western Islands circa 1695,* 2nd edition. Facsimile of 2nd edition 1981, Edinburgh: Mercat Press..

Nelson, E.C. (1983). Tropical drift fruits and seeds on coasts in the British Isles and Western Europe II. History (1560-*c*.1860) and Folk-lore. *Scottish Naturalist* 1983, 11-63.

Radford, E. & Radford, M.A. (1969). *Encylopaedia of Superstitions,* pp. 64-65. London: Hutchinson.

James Robertson, an Eighteenth Century Discoverer of Scotland

D.M. HENDERSON

Larachan, 54 Lonemore, Gairloch, Ross-shire, IV21 2DB

Summary

A short note is presented drawing attention to the contribution James Robertson made to studies on 18th century Scotland and to early knowledge of the flora of the country.

In the history of the discovery of Scotland there has always been a shadowy figure of one 'James Robertson'. Some scattered details of him were known but persistent searching and a large measure of good luck has unearthed a great story of one of our earliest and most successful travellers in Scotland (Henderson & Dickson, 1994).

James Robertson was born about 1740, the nephew of John Williamson, head gardener to John Hope, Professor of Botany and Keeper of the Botanic Garden in Leith Walk, Edinburgh. Hope trained him and sent him exploring the natural history of Scotland. His reports to the Commissioners of the Annexed Estates for the five journeys they funded provide detailed accounts of his observations throughout most of Northern Scotland and the Western and Northern Isles. These precede Lightfoot's journeys with Thomas Pennant. Further manuscripts reveal tabular accounts of the flora and fauna of the Islands of Scotland. Underlying the facts there is an undercurrent of secrecy by Robertson and Hope to prevent their information being pirated by others. The papers present a remarkable assemblage of observations on agricultural improvement, concern for drifting sands, folk medicine, the possibilities for a Caledonian Canal, and the fishing industry in the Northern Isles. But their greatest contribution is to the early knowledge of the flora of Scotland, adding many first finds for the British Isles. After 1771 Robertson travelled to China and became a doctor with the East India Company, eventually returning to Fife, a moderately rich man, and dying there in 1796. His illegitimate son and only child migrated to America and his direct descendents have provided much invaluable information on Robertson's later life.

Reference

Henderson D.M. & Dickson, J.H. (1994). *A Naturalist in the Highlands. James Robertson, his Life and Travels, 1767-1771,* 262pp. Edinburgh: Scottish Academic Press.

The Gathering of Botanical Information in Ireland, 1600 to 1800

E. CHARLES NELSON

National Botanic Gardens, Glasnevin, Dublin 9, Republic of Ireland

Summary

Botanical exploration in Ireland before 1800 was conducted mainly by individuals, *e.g.* Richard Heaton (*c.* 1640), William Sherard (*c.* 1692), Caleb Threlkeld (*c.* 1726). There was one attempt in the mid-1700s by the Physico-Historical Society to organize a concerted programme of botanical investigation. This paper concentrates on three separate recorders of the Irish flora and their extant manuscripts, Don Philip O'Sullivan Beare (*c.* 1625), John Rutty (*c.* 1740) and Patrick Browne (*c.* 1780). A manuscript account of the Irish flora survives from the first quarter of the seventeenth century. Philip O'Sullivan Beare wrote *Zoilomastix* about 1625, as a response to the disparaging writings of such authors as Giraldus Cambrensis. O'Sullivan Beare clearly had first-hand knowledge of some gardens and forests, and seems to have been well-versed in botany. The Physico-Historical Society of Ireland promoted natural history exploration. Dr John Rutty was the person who apparently co-ordinated the programme. Isaac Butler carried out the field work. Butler's records occur in annotations made by Rutty in a copy of Threlkeld's *Synopsis Stirpium Hibernicarum.* Linnaeus' work did not find a keen disciple in Ireland until the return home from the West Indies of Patrick Browne whose *The Civil and Natural History of Jamaica* (1756) was one of the first post-Linnaean floristic works. Browne observed and collected plants in Ireland after 1770 and his unpublished manuscript contains valuable information about Irish vernacular names.

Introduction

'It has been justly considered a reproach to Ireland that the neighbouring kingdoms know more of the indigenous plants and seeds of almost every part of the habitable globe, and of most of the little islands, which have been discovered in the South Sea, than of this country distinguished in all other respects for scientific attainments and the encouragement of liberal arts.' (J. Hely Hutchison, n.d.)

Hutchison's comparison, in the late eighteenth century, of knowledge of the floras of other lands was an ingenious allegory but not a valid one - he was not comparing like with like, nor was he well-informed. Ireland was, relatively, a botanical backwater because of its isolated geographical situation, and the social, economic, and intellectual history of the country, but she was by no means ill-served by botanists. Botanical exploration of Ireland before 1800 was mainly the preserve of individual enthusiasts, 'Lovers of Botanick Exercises', to apply a description employed by Rev Dr Caleb Threlkeld (1726), who 'used to perambulate ... into the Fields and Fells ... to have ocular demonstration of the Plants themselves in their native Soil'. However, there was one planned programme of natural history exploration, including botany, during the mid-eighteenth century, involving a 'professional' botanist.

Given that botanical exploration was a solitary occupation, compilation of a cadastre of the Irish flora was largely a haphazard process: before 1800 only one flora had been published, county lists were few and usually sparse, and only the 1794 Dublin catalogue by Walter Wade employed binary (Linnaean) names. While

Irish-based botanists were not full-time academics nor even, as in the nineteenth century, professional horticulturists with strong botanical interests, by 1800 more than half of the native higher plants had been catalogued, and considerable progress had been made to record cryptogams although major studies of lower plants were not commenced until the early nineteenth century.

Before 1600

There are few ways of gauging knowledge of plants in Ireland before 1600. Certainly plants were cultivated for food, and were collected by herbalists, physicians and craftsmen for use, for example, in medicines and as sources of dyes. A number of mediaeval medical manuscripts survive but invariably these are copies of 'classical' texts without indigenous information except Irish vernacular names. Literary works by native Irish poets also contain many references to plants, but none can be regarded as scientific texts; like the innumerable references to wild plants in Shakespeare's plays, Irish poems and sagas merely indicate that, as one would expect, plants were integral to everyday life.

There are no references to plants from Ireland in herbals published in Britain before 1600; Gerard (1597) alone noted that the Irish used the word shamrock for white and red clovers (*Trifolium repens* L., *T. pratense* L.; see Nelson, 1991).

Philip O'Sullivan Beare

The first original author to record Irish plants in a systematic manner was an exile, Philip O'Sullivan Beare, who about 1625 compiled '*Zoilomastix*' as a retort (*cf.* O'Donnell, 1960) to such dubious tracts as Giraldus Cambrensis's *Topographia Hibernia*, and Stanihurst's *De rebus in Hiberniae gestis*. O'Sullivan Beare, born in County Cork but exiled in Spain from boyhood, managed to assemble a corpus of information that indicates substantial knowledge about his homeland that was not merely hearsay. '*Zoilomastix*' survives in manuscript (photostat copy in National Library of Ireland, Dublin: MS 2759-62); chapters on history, religious matters and topography have been published but the botanical chapters (comprising 17 ff.), while studied for Irish names, have not been published. Chapter 42 concerns *Viventia, quae sensu carent* ('Living things that are devoid of feeling'):

> 'There remain other living things ... trees, shrubs, grasses, with which that island is abundantly supplied. ... it would be a lengthy business to discover the names of each of them and more than one writer would need to be consulted. But, lest we should move away without sampling any of the copious delights from such a noteworthy store, I thought of beginning with some words regarding fruit-bearing trees.'

O'Sullivan Beare noted Latin, Greek, Irish and Spanish names for each plant, with a commentary which was sometimes evidently original but frequently quoted such authors as Pliny, Calepino and Dalecamp. Of *Arbutus unedo* L., which he must have seen in Spain, and evidently knew in Ireland, O'Sullivan Beare wrote:

> 'The wild-strawberry tree produces a fruit the Greeks call *memecylon*. The arbutus produces a fruit called *unedo* ... hence because (it bears) only a single (fruit) from which many more are produced, red in colour and soft.'

Other passages indicate his most intimate knowledge of Ireland: quince 'was planted and bore fruit in Connacht in the garden of John Burke; it was planted by a medical man'.

Other seventeenth century botanists

Parkinson (1640) reported that *Arbutus* has been 'of late dayes found in the west part of Ireland' - and also that sundew (*Drosera* sp.) was found in Ireland. Thus to

Parkinson must go credit for the earliest explicit references *in print* to native Irish flowering plants (Mitchell, 1975). A decade later, How (1650) added *Dryas octopetala* L. and *Gentiana verna* L. to the embryonic inventory of Ireland's flora (*cf.* Nelson & Walsh, 1991), as well as *Juniperus communis* L., *Rubus saxatilis* L. and *Epipactis* sp.; his informant was the vicar of Birr, the Yorkshire divine, Rev Richard Heaton (Walsh, 1978).

In the latter half of the seventeenth century there was a flurry of itinerant botanists beginning with the Welsh antiquarian, Edward Lhuyd, who visited Ireland on two occasions, and made extensive journeys into Tipperary (Nelson, 1993a) and the north and west as far as Kerry. Lhuyd was interested in plants, antiquities and the Irish language, and his gatherings of plants survive. On Lhuyd's second visit he discovered some of the choicest native species including *Daboecia cantabrica* (Hudson) K. Koch (illustrated in Petiver, 1703: '*mulierculae superstitiosae surculos ejus secum circumferunt adversus incontinentiam*') in Connemara (Nelson, 1979b), and *Adiantum capillus-veneris* L. on the Aran Islands.

William Sherard lived in Ireland for about three years between the late summer of 1690 and end of 1693 and collected in County Down including the Mourne Mountains, and visited Dublin (Nelson, 1983), while Gédéon Bonivert, a Dutch officer in the army of William of Orange about the same time, gathered plants in Cork (Wijnands, 1991). None of these travellers provided accounts of the substance of that of O'Sullivan Beare, but at least the records of Bonivert and Sherard were published by British authors including Ray and Plukenet, and Lhuyd published his own records in *Philosophical Transactions* (Mitchell, 1975).

Early eighteenth century

The principal workers in the first half of the eighteenth century were two resident clerics: Caleb Threlkeld published *Synopsis Stirpium Hibernicarum* ... in October 1726 (Nelson, 1979a; 1993b; Nelson & Synnott, 1988), and his unreliable contemporary, John Keogh, chaplain in the household of the Earl of Kingston, published a 'useless' ([Baily] 1833) herbal in Cork (Keogh, 1735). Threlkeld was assisted by Dr Thomas Molyneux, Professor of Physic in Trinity College Dublin. Molyneux, one of the several graduates of the University of Leiden who were based in Dublin during the late seventeenth and early eigheenth centuries, is an enigmatic character because, while he was clearly interested in plants, he does not appear to have stimulated botanical work. In 'A discourse concerning the large horns frequently found under ground in Ireland', Molyneux (1697) reported *Saxifraga spathularis* Brot. from County Kerry. His older brother, William, founded the Dublin Philosophical Society in 1683 - that society had heard botanical papers such as an account of regeneration of poppies (Hoppen, 1970).

Dr John Rutty, also a Dublin-based, Leiden-trained physician, was a Wiltshire Quaker whose remarkable diary contains such agonized entries as

> 'Seventh month 1764: 30. Still cause of complaint of earthly attractions, even to natural studies ... Farewell now plants, farewell fossils and waters, any more than as in the way of duty, and without anxiety; for God hath placed higher and better objects before me.
>
> Fifth month 1769: 29. The botanick walk: O the perverseness, the indolence, the ignorance and the sin of the apothecaries'. (Rutty, 1796: pp. 276, 379).

For two decades (*c.* 1724-1744) after he established himself in Dublin, Rutty noted, in the margins of a copy of Threlkeld's *Synopsis* ... (Nelson, 1990; 1992b), plants that he and others (*e.g.* Dr Charles Lucas (*cf.* Nelson, 1992a)) had observed; Rutty's localities range from Dublin to Carlow in the south-east and Cavan in the north-west. He recorded this period of his life as follows:

> 'From 1740 to 1745, engaged in the Natural History of the County of Dublin, and was led a merry dance on birds, fishes, and fossils, and in compotations for information, and was greatly hurt in my spirituals by these means, preferring nature to grace, and moral righteousness to evangelical'. (Rutty, 1796: p. 362).

Rutty is often dismissed as an eccentric and has not received due credit for his natural history studies, especially for his activities within the Physico-Historical Society of Ireland. This was formed on 14 April 1744 and had as its prime purpose a county by county survey of Ireland; the last meeting recorded in the society's minute-book was held on 27 March 1752 (Davies, 1979). The society employed Isaac Butler (1689-1755) to carry out surveys and collect botanical and geological specimens. On 28 May 1744, the minute-book (Royal Irish Academy, Dublin: MS 24 E 28) records

> 'Ordered that the Treasurer (or in his absence ye Secretary) do pay into the hands of Mr Harris one guinea for the use of Isaac Butler Botanist in order to encourage him in Inquiries after rare plants &c in ye County of Dublin'.

One week later 3 guineas were sanctioned for Butler to travel 'the north part of the County of Dublin, the County's of Meath and West-Meath in search of simples, fossils and such other curiosities as are mentioned in the Society's proposals'. At the society's July 1744 meeting, Butler's territory was extended into the counties Louth and Longford, and he was to be paid 2 guineas 'for his additional troubles.' The results were sufficiently interesting for Rutty to produce a catalogue of plants and minerals collected by Butler at the August 1744 meeting, but by October Butler was ill, and the society paid a guinea for his support. In November 1745, Dr Rutty 'produced specimens and accounts of several natural curiosities collected thro' the counties of Dublin ... [including] An account of rare plants, near Lucan.' At the January 1746 meeting, Rutty again displayed specimens from County Dublin. More than a year later, after a lull in the society's botanical activities, three members, Harris, Rutty and Joseph Butler, were instructed to 'agree with Mr Isaac Butler Botanist for a journey in search of simples & thro' the County of Antrim' but the proposal was amended ten days later and Butler's proposed survey was altered to County Cork. However Butler, when approached, was unable to undertake any expedition that summer of 1747 - he was apparently engaged on 'taking the circumference in gross of the City of Dublin.' In February 1748, a resolution was passed as follows:

> 'It appears to the committee that there was an order to ye General Board on 10th June 1747 that Mr Isaac Butler be sent down to ye Co. of Cork to assist Mr [Charles] Smith in ye Natural Hist. of ye said Country wherefore it is the opinion of the Committee that Dr Rutty do agree with the said Butler for such a journey at any sum not exceeding 3/6 sterling per diem and also that it be an instruction to Dr Rutty that he direct the said Butler to pursue his enquiries into the adjacent counties of Limerick, Kerry or Tipperary as he can have time'.

Butler accepted these terms, and in December following, Rutty produced a 'list and several specimens of plants and other curious particulars of Natural History collected by Isaac Butler in his late journey thro' the counties of Cork, Tipperary, Wexford, Wicklow etc.'

The results of the Physico-Historical Society's surveys were not published formally in periodicals or books, but the data garnered were available to members for use in publications during the society's active life and after its cessation, principally Harris (1744) and Smith (1746; 1750; 1756). Rutty's natural history of Dublin was delayed until 1772 by which time the society was extinct and that work was 'accommodated to the noble designs of the Dublin Society': at least in Rutty's book (1772), Butler's activities are acknowledged with all the plants he observed indicated.

When compared with the prevailing situation of botany in Britain, the Physico-Historical Society's work is most remarkable. During the mid-1700s natural history in Britain 'experienced one of its thankfully rare breakdowns in elementary manpower' (Allen, 1976), formal societies were moribund, and in botany there was relative torpor. No British society was employing collectors on the same terms as Butler.

Linnaeus and Irish plants

The Physico-Historical Society ceased to meet in 1752, one year before *Species Plantarum* was published. No copies of the first edition of this seminal work are in Irish libraries today (Nelson, 1981), suggesting that the works of Linnaeus had no receptive audience in Ireland. Comments by later botanists, particularly Patrick Browne and Robert Brown, about the dearth of late eighteenth century botanical texts in Ireland, tend to support this view.

Rutty (1772) acknowledged that Dr Abraham Lionel Jenkins, a member of the Physico-Historical Society, had 'during a long series of years has been preparing a catalogue of native Irish plants, adjusting their names to the Linnaean system' but no trace of Jenkins' catalogue survives. Jenkins assisted Harris in the preparation of his account of County Down (Harris, 1744) and is said also to have written various essays on Irish natural history.

Smith's *The Antient and Present State of the County of Kerry* (1756) was the first work containing botanical lists issued in Ireland after the publication of *Species Plantarum.* Rutty's *An Essay towards a Natural History of the County of Dublin* (1772) was the next but neither Smith nor Rutty included names in binary form. Thus Wade's Dublin list (1794) was the first Irish publication to follow Linnaeus' nomenclatural scheme.

The earliest extant catalogues of Ireland's fauna and flora with Linnaean names were compiled by Dr Patrick Browne (Nelson & Walsh, 1994); lists of birds and fishes (including marine mammals) were published in 1774, but his *'Fasciculus Plantarum Hiberniae'* (Ms. in Linnean Society, London) has not yet been published.

Browne, a native of Mayo, studied medicine in Paris between 1738 and 1742, graduated at the University of Rheims on 1 December 1742, and matriculated at Leiden in February 1743, before settling in London as a physician in St Thomas's Hospital. By 1746, Browne had returned to the West Indies and ten years later he published *The Civil and Natural History of Jamaica*, illustrated by Georg Ehret, in which he adopted Linnaeus' system of classification (Browne, 1756). Browne retired from active medical duties in 1770 and returned to Ireland; on 7 September 1770, in a letter to Linnaeus he outlined his plans:

> 'It is now some years since I had ye pleasure of hearing from you ... I am ... returned to this Kingdom, my native country, where I hope to pass the remainder of my days ... I have been already a considerable distance thro ye country & have seen only 2 plants that I could not recolect any thing of, ye one is a species of Anthericum, growing frequently in our bogs, ye other intirely unknown to me ...

the last edition of your works is not to be had in this Kingdom & if you could find means to send me a sett of your *Systema Genera, & Species*, by any vessell from Stockholm, I would be greatly Oblidged to you ...'. (Hulth, 1916: 355-356).

Browne's manuscript flora, dated 30 August 1788, included a preface that is a manifesto for Irish botany and botanic gardens (*cf.* Nelson & McCracken, 1987), and listed approximately 600 vascular plants, and about 300 cryptogams. When Browne was unable to equate a plant with one described and named by Linnaeus, he wrote his own description and suggested a binomial. As Browne lived in retirement at Rushbrook, a townland south of Claremorris, his catalogue includes localities for plants that abound in the peatlands, and also those common on the limestone outcroppings in south-eastern Mayo. Browne also visited The Reek (Croagh Patrick) recording about 20 species thereabouts.

Conclusion

The successors of Philip O'Sullivan Beare and Patrick Browne included Walter Wade, Robert Brown, John Templeton and James Townsend Mackay. Wade published the first Linnaean flora, while Brown and Templeton explored the cryptogamic flora with enthusiasm but without publishing. Mackay issued a series of papers (Mackay, 1806; 1807) followed by a catalogue of Irish native plants (Mackay, 1825), and lastly *Flora Hibernica* (Mackay, 1836) which remains the only flora of Ireland to contain both phanerogams and cryptogams. Meanwhile Baily's anonymous flora (1833) had this echoing preface:

'It is a matter of surprise that Ireland should have remained so long without any descriptive catalogue of its plants, worthy of notice. It has been remarked that when England and France had their provincial Flora, the Botany of this island was as much unknown as that of any island in the Pacific.'

Acknowledgements

I am especially grateful to Mary Brennan for her translations of Latin texts, especially of O'Sullivan Beare's '*Zoilomastix*', used herein. My thanks are also due to the Royal Irish Academy, Dublin, the Department of Manuscripts, Trinity College, Dublin, the Linnean Society, London, and the National Library of Ireland, Dublin, for access to manuscripts.

References

Allen, D.E. (1976). *The naturalist in Britain. A social history*, xii, 292 pp. London: Allen Lane.

[Baily, K.S.] (1833). *The Irish flora; comprising the phaenogamous plants and ferns*, xii, 220 pp. Dublin: Hodges & Smith.

Davies, G.L.H. (1979). The making of Irish geography, IV: The Physico-Historical Society of Ireland, 1744-1752. *Irish Geography* **12**, 92-98.

Gerard, J. (1597). *The Herball or Generall Historie of Plantes*, xviii, 1392, 72 pp. London: John Norton.

Harris, W. (1744). *The Antient and Present State of the County of Down*, xx, 292 pp. Dublin.

Hoppen, K.T. (1970). *The Common Scientist in the Seventeenth Century. A Study of the Dublin Philosophical Society 1683-1708*, xiv, 297 pp. London: Routledge & Kegan Paul.

[How, W.] (1650). *Phytologia Britannica*, xvi, 134 pp. London: Octavian Pulleyn.

Hulth, J.M. (1916). *Bref och Skrifvelser af och till Carl von Linne*. ser. 2(1). Stockholm.

Hutchison, J.H. (n.d. [before 1794]). 'An essay towards a history of Trinity College Dublin'; unpublished (Trinity College, Dublin: MSS. 1770-1774b; Book II chapter 4).

Keogh, J. (1735). *Botanalogia Universalis Hibernica*, vi, xii, 146, 32 pp. Cork: George Harrison.

Mackay, J.T. (1806). A systematic catalogue of rare plants found in Ireland. *Transactions of the Dublin Society* **5**, 121-184.

Mackay, J.T. (1807). A systematic catalogue of rare plants found in Ireland. *The Dublin Medical and Physical Essays* **1**, 328-343; **2**, 106-117.

Mackay, J.T. (1825). A catalogue of the plants found in Ireland. *Transactions of the Royal Irish Academy* **14**, 103-198 (also issued separately as *A Catalogue of the Plants found in Ireland, with Descriptions of some of the Rarer Sorts. Part I*, 98 pp. Dublin: Royal Irish Academy).

Mackay, J.T. (1836). *Flora Hibernica, Comprising the Flowering Plants and Ferns Characeae Musci Hepaticae Lichenes and Algae of Ireland*, xxxviii, 271 + 279 pp. Dublin: William Curry jun.

Mitchell, M.E. (1975). Irish botany in the seventeenth century. *Proceedings of the Royal Irish Academy* **75B** (13), 275-284.

Molyneux, T. (1697). A discourse concerning the large horns frequently found under ground in Ireland. *Philosophical Transactions of the Royal Society* **19**, 489-512.

Nelson, E.C. (1979a). 'In the Contemplation of Vegetables' - Caleb Threlkeld (1676-1728), his life, background and contribution to Irish botany. *Journal of the Society for the Bibliography of Natural History* **9**, 257-273.

Nelson, E.C. (1979b). Historical records of the Irish Ericaceae, with particular reference to the discovery and naming of *Erica mackaiana* Bab. *Journal of the Society for the Bibliography of Natural History* **9**, 289-299.

Nelson, E.C. [1981]. Works of botanical interest published before 1800 held in Irish libraries. *National Botanic Gardens, Glasnevin, Occasional papers* **1**, 127 pp.

Nelson, E.C. (1983). Sir Arthur Rawdon (1662-1695) of Moira: his life and letters, family and friends, and his Jamaican plants. *Proceedings of the Belfast Natural History and Philosophical Society* **10**, 30-52.

Nelson, E.C. (1990). 'It's a long way to Tipperary' - natural history archives in Ireland with an appendix listing archives. *Archives of Natural History* **17**, 325-347.

Nelson, E.C. (1991). *Shamrock. Botany and History of an Irish Myth*, xiv, 200 pp. Aberystwyth & Kilkenny: Boethius Press.

Nelson, E.C. (1992a). Charles Lucas' letter (1736) to Sir Hans Sloane about the natural history of The Burren, County Clare. *Journal of the Irish Colleges of Physicians and Surgeons* **21**, 126-131.

Nelson, E.C. (1992b). 'Out of old bokes ... cometh all this newe science': personal reflections on botany, botanists and books. *BSBI News* **62**, 24-31.

Nelson, E.C. (1993a). Botany Mountain, Ireland: 1690s-1990s: or, Edward Lhuyd and an Irish myth. *BSBI News* **63**, 21-22.

Nelson, E.C. (1993b). Searching the archives for botanists, with some Irish case histories. *Huntia* **9**, 5-19.

Nelson, E.C. & McCracken, E.M. (1987). *'The Brightest Jewel': a History of the National Botanic Gardens, Glasnevin, Dublin*, [xiv], 275 pp. Kilkenny: Boethius Press.

Nelson, E.C. & Synnott, D.M. (1988). *The First Irish Flora: Caleb Threlkeld's Synopsis Stirpium Hibernicarum* (1726), liv, 176, 60, 48 pp. Kilkenny: Boethius Press.

Nelson, E.C. & Walsh, W.F. (1991). *The Burren: a Companion to the Wildflowers of an Irish Limestone Wilderness*, [vii], 343 pp. [Aberystwyth & Ennis]: Boethius Press & The Conservancy of The Burren.

Nelson, E.C. & Walsh, W.F. (1994). *Flowers of Mayo. Dr Patrick Browne's Fasciculus Plantarum Hiberniae 1788.* Dublin: De Burca Rare Books. (in preparation).

O'Donnell, T.J. (1960). *Selections from the Zoilomastix of Philip O'Sullivan Beare.* Dublin.

Parkinson, J. (1640). *Theatrum Botanicum*, xvii, 1745, 3 pp. London. Publisher: T. Cotes.

Petiver, J. (1703). *Gazophylacii Naturae et Artis.* London.

Rutty, J. (1772). *An Essay towards a Natural History of the County of Dublin*, xii, 392 + vi, 488 pp. Dublin: the author.

Rutty, J. (1796). *Spiritual Diary, and Soliloquies*, xx, 421 pp. London: James Phillips. 2nd edition.

Smith, C. (1746). *The Antient and Present State of the County and City of Waterford*, 386 pp. Dublin.

Smith, C. (1750). *The Antient and Present State of the County and City of Cork*, 434 + 436 pp. Dublin.

Smith, C. (1756). *The Antient and Present State of the County of Kerry*, 424 pp. Dublin.

Threlkeld, C. (1726). *Synopsis Stirpium Hibernicarum*, 26, 176, 60 pp. Dublin: the author. (Facsimile edition: see Nelson & Synnott, 1988).

Wade, W. (1794). *Catalogus Systematicus Plantarum Indigenarum in Comitatu Dublinensi Inventarum. Pars prima*, vii, 275, 31 pp. Dublin: Dublin Society.

Walsh, L. (1978). *Richard Heaton of Ballyskenagh 1601-1666*, 114 pp. Roscrea: Parkmore Press.

Wijnands, D.O. (1991). Plants collected by Gédéon Bonivert (1651-1703) in the garden of Johan Stickers (1630-1701). *Archives of Natural History* **18**, 27-29.

Linnaeus as an Economic Botanist

WILLIAM T. STEARN

Honorary Research Associate, Royal Botanic Gardens, Kew, Richmond, Surrey, TW9 3AE

Summary

The interest of Carl Linnaeus (1707-1778) in the uses of plants as manifested in his travel books and dissertations was much influenced by the urge of his country, war-impoverished Sweden, for national recovery and self-sufficiency. His official Swedish journeys of 1741-1749 were primarily economic surveys, giving much attention to locally used dye-plants. The dissertation *Pan Suecicus* (1749) details experimental feeding of cattle, pigs, sheep and goats with native Swedish plants: it also represents an important stage in the development of the Linnaean binomial method for naming species.

A massive volume published in 1907 by the Royal Swedish Academy of Sciences to commemorate the two-hundredth anniversary of the birth of its founder member, the Swedish naturalist Carl Linnaeus (1707-1778), after 1762 Carl von Linné, contains treatises on him as a medical man, as a zoologist, as a botanist, as a geologist and as a mineralogist. Other publications have lauded him as dietician, paediatrician, horticulturist, pioneer biochronologist, mountain ecologist, also rather surprisingly as artist, less so as sexualist. He can also be equally esteemed as a pioneer economic botanist and thus is a fitting subject for this Edinburgh symposium on 'Plants and People'.

Linnaeus's interest in uses of plants was much influenced by the domestic economy of the Sweden of his times, a war-weary and war-impoverished defeated country intent, however, upon recovery, development and self-sufficiency after the disastrous wars of its valiant King Charles XII (1682-1718). Gustavus Adolphus and his able chancellor Axel Oxenstierna had made Sweden by 1648 a great military power with territory extending over most of the Baltic region except Denmark. This included the important grainlands of Livonia (part of Latvia). During the reign of Charles XII and soon afterwards Sweden lost most of this territory, having aroused the envy, hostility and land-acquisitiveness of other northern countries. It became almost poverty-stricken. A political party, the Hats, with strongly mercantilist policies, then rose to power in the Swedish parliament and ruled until 1765. Much to his advantage Linnaeus became associated with the Hats. Their support enabled him to make his journeys of economic survey on Öland and Gotland, in Västergötland and Skåne.

When Linnaeus returned to Sweden in 1738 after his productive stay in Holland, where nine books had earned him renown, he found himself in Stockholm to be simply an unimportant physician. However he had early learned the value of self-publicity and self-esteem to impress influential people with his talents. Soon he came to the notice of the cultured and powerful Count Carl Gustaf Tessin (1695-1770) who in 1739 became leader of the Hats party and introduced him to Stockholm society. In 1739 six men, Linnaeus being one, founded the Royal Swedish Academy of Science. Its aims included the study of mathematics, natural history, economics, trade, useful arts and manufactures. Thus the infant Academy began with a marked utilitarian purpose in accordance with the country's needs. The investigation of Sweden's natural resources, of special interest for Linnaeus, came well within the Academy's scope. As stated by Nelson and Svanberg (1987), the travels of Linnaeus were largely expeditions to collect folk knowledge in the countryside which could benefit the nation as a whole.

In 1741 Linnaeus started such an investigation. Having been appointed professor of medicine and botany at the university of Uppsala, he set out on 15 May 1741, with six companions, officially to survey the natural resources of the then relatively little known Baltic islands of Öland and Gotland, which, being of limestone, are very different from the granitic Swedish mainland. To quote Linnaeus's own statement: 'In the Riksdag of 1741 the Estates were pleased to decree that I should make a journey to several districts of the Kingdom, especially to Gotland, Öland, Västergötland ....'. One can easily surmise who instigated that decree. Among his instructions were: 'to search for plants suitable for making dyes; to investigate plants, hitherto imported for pharmaceutical purposes, which could be found within the Kingdom'. The group returned from their long journey on 28 August 1741, having zealously and successfully obeyed all these and other instructions. Linnaeus later published their collective observations in his *Öländska och Gothländska Resa* (1745).

This interesting travel-book, many times reprinted, contains information not only about animals and plants here recorded for the first time in Sweden but also about wedding customs, household remedies, rune-stone inscriptions, fences, crops and much else. Linnaeus's enquiries revealed that the country folk, like peasants elsewhere, used a diversity of plants for dyeing, sometimes the same plant for more than one colour (cf. Hofmann, 1992). They obtained a **red** dye from *Origanum vulgare*, roots of *Galium triandrum* and a lichen (*Ochroleucia tartarea*): a **yellow** dye from *Betula* leaves, *Anthemis tinctoria, Reseda luteola, Bidens cernua, Serratula tinctoria* and a lichen (*Cetraria juniperina*); a **brown** dye from bark of *Rhamnus catharticus, Origanum vulgare* and a lichen (*Parmelia saxatilis*); an **orange** dye from *Bidens tripartita;* a ***green*** dye from *Frangula alnus* and *Succisa pratensis;* a **violet** dye from berries of *Sambucus ebulus*. Linnaeus's observations are the first on dye-producing plants of Sweden.

In addition to his *Materia medica* Vol. 1 *De Plantis* (1749) Linnaeus published many dissertations on medicinal and other economic plants. These pamphlets bear the names of students who nowadays would be claimed as their authors but Linnaeus was wholly or partly responsible for their content. They were produced according to an academic procedure of mediaeval origin accepted at Uppsala until 1852 and also during the 18th century and later at the universities of Halle, Helmstädt, Jena, Lund, Tübingen and Wittenberg (*cf.* Horn, 1893). At these universities the professor drafted the thesis for a degree; its defence in a public disputation tested the student's proficiency in logic and spoken Latin. The following dissertations attest Linnaeus's long-continued interest in economic botany: *Flora oeconomica* (1748), *Radix Senega* (1749), *Pan Suecicus* (1749), *Plantae esculentae Patriae* (1752), *S Rhabarbarum* (1752), *Plantae officinales* (1751), *Spigelia Anthelmia* (1758), *Plantae tinctoriae* (1759), *Inebrantia* (1763), *Lignum Quassiae* (1763), *Hortus culinaris* (1764), *Purgantia indigena* (1766), *Menthae Usus* (1767), *Violoa Ipecacuanha* (1774) and others.

The most interesting of these is *Pan Suecicus* (1749), defended by N.L. Hesselgren (Fig. 1). This records the experimental feeding of livestock with about 800 species of Swedish plants. Its origin goes back to Linnaeus's Lapland journey of 1732. He noticed that his horse ate only certain plants out of the diversity available and avoided eating those known to us as poisonous, *e.g. Aconitum, Convallaria.* In his *Flora Lapponica* (1737) he suggested that it would be useful to ascertain which plants were palatable, which noxious to livestock but not until 1747 and 1748 did he have the opportunity of doing so experimentally. Nine persons, among them Hesselgren, took part in this enquiry. They gave plants freshly gathered to cattle, goats, sheep, horses and pigs, when they would not be excessively hungry; some experiments were repeated up to 10 times. Linnaeus set out the results systematically and concisely with six columns in *Pan Suecicus,* each plant being designated in the lefthand column not by a lengthy descriptive phrase-name but simply by its *Flora Suecica* (1746) number, its generic name and an epithet, *e.g. 50 Phleum vulgare* instead of *Phleum spica cylindrica longissima.* The other columns were headed B *(Boves,* cattle*)*, C *(Caprae,* goats*)*, O *(Oves,* sheep*)*, E *(Equi,* horses*)* and *S (Sues,* pigs*)*. Acceptance of a particular plant is indicated by I, neglect by 'O', not given by '-'. Linnaeus's summary states that: **Cattle** ate 276, neglected 218 plants; **Goats** ate 449, neglected 336; **Sheep** ate 387, neglected 141; **Horses** ate 262, neglected 212; **Pigs** ate 72, neglected 171. This pioneer exercise in economic botany and animal husbandry aroused much interest at the time and Richard Pulteney published an English translation in *The Gentleman's Magazine* Vol. 28 (1758) and his *General View of the Writings of Linnaeus* (1758). It seems not to have been repeated so comprehensively and merits a modern agricultural and veterinary study.

These two works *Öländska och Gothländska Resa* and *Pan Suecicus,* are also historically important in foreshadowing Linnaeus's consistent adoption of binomials for all the world's plants then known in his *Species Plantarum* (1753) which has made this work the internationally accepted starting point for modern botanical nomenclature. He first explained his method in *Pan Suecicus*: 'If we take the *Flora Suecica,* Stockholm, 1745, and for each herb, *in order to save paper* [italics mine], we put the generic name, the number of the *Flora Suecica* in place of the differential diagnosis, the matter is easily put in handy form'. For example, *Geranium pedunculis bifloris, calycibus pyramidatis angulatis, foliis quinquelobis rotundatis,* designating a species found on Gotland and thus described in *Öländska och Gothländska Resa,* was indexed in that work as *Geranium* 574 *lucidum,* a name retained, as *Geranium lucidum,* in *Species Plantarum.* Thus, as stated elsewhere, 'The genesis of the Linnaean binary system of nomenclature is to be found in the index to Linnaeus's *Öländska och Gothländska Resa* (1745) its first trial application in the dissertations *Gemmae Arborum* (November 1749) and *Pan Suecicus* (December 1749), its first enunciation as a method of naming in *Philosophia botanica* (1751), its first application to the vegetable kingdom as a whole in *Species Plantarum* (1753). It originated as an indexer's paper-saving device, then proved convenient for naming both plants and animals in general' (Stearn, 1957). That surely is the most long-lasting and biologically the most important outcome of Linnaeus's 1741 journey of economic survey.

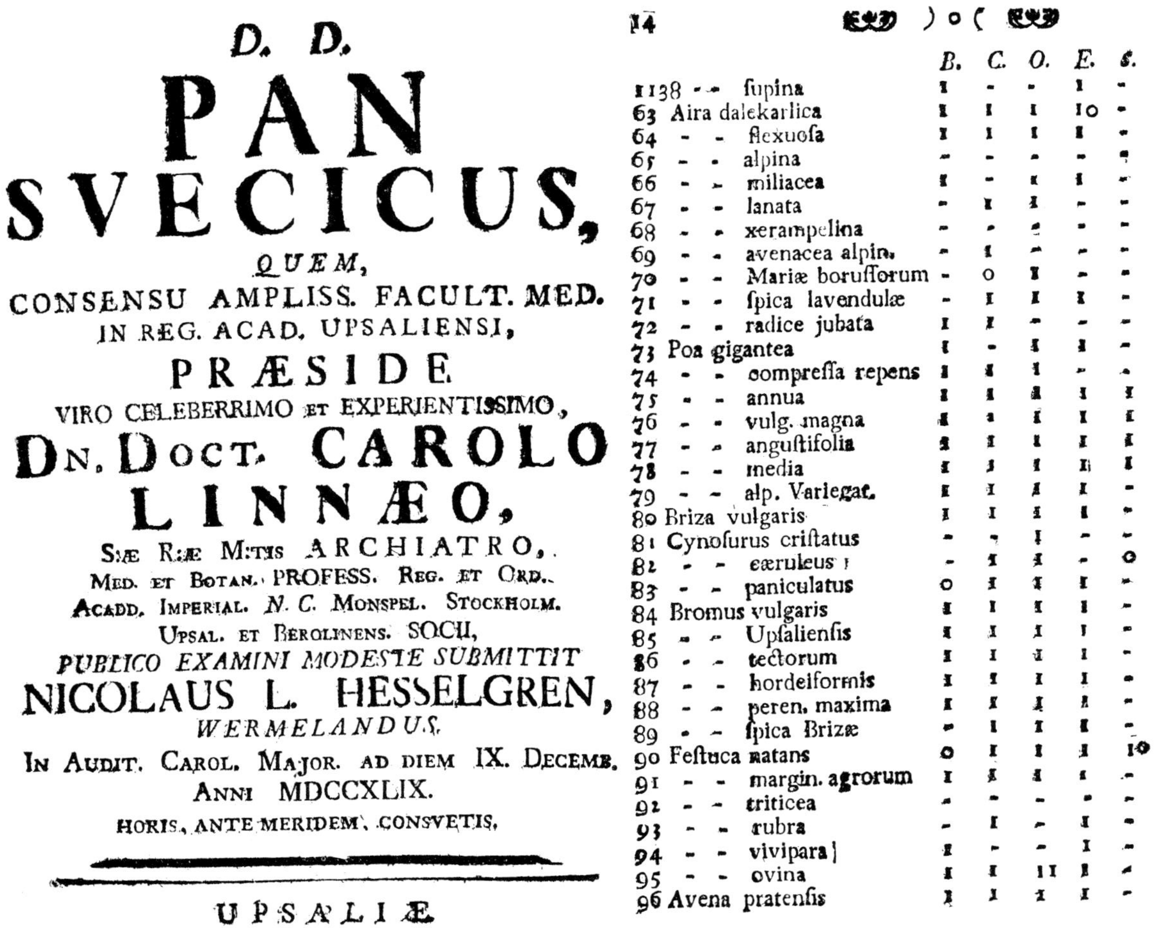

D. D.
PAN SVECICUS,
QUEM,
CONSENSU AMPLISS. FACULT. MED.
IN REG. ACAD. UPSALIENSI,
PRÆSIDE
VIRO CELEBERRIMO ET EXPERIENTISSIMO,
DN. DOCT. CAROLO
LINNÆO,
S:Æ R:Æ M:TIS ARCHIATRO,
MED. ET BOTAN. PROFESS. REG. ET ORD.
ACADD. IMPERIAL. N. C. MONSPEL. STOCKHOLM.
UPSAL. ET BEROLINENS. SOCII,
PUBLICO EXAMINI MODESTE SUBMITTIT
NICOLAUS L. HESSELGREN,
WERMELANDUS.
IN AUDIT. CAROL. MAJOR. AD DIEM IX. DECEMB.
ANNI MDCCXLIX.
HORIS ANTE MERIDEM CONSVETIS.

UPSALIÆ

14

		B.	C.	O.	E.	S.
1138	- - fupina	1	-	-	1	-
63	Aira dalekarlica	1	1	1	10	-
64	- - flexuofa	1	1	1	1	-
65	- - alpina	-	-	-	-	-
66	- - miliacea	1	-	1	1	-
67	- - lanata	-	1	1	-	-
68	- - xerampelina	-	-	-	-	-
69	- - avenacea alpin.	-	1	-	-	-
70	- - Mariæ borufforum	-	o	1	-	-
71	- - fpica lavendulæ	-	1	1	1	-
72	- - radice jubata	1	1	-	-	-
73	Poa gigantea	1	-	1	1	-
74	- - compreffa repens	1	1	1	-	-
75	- - annua	1	1	1	1	1
76	- - vulg. magna	1	1	1	1	1
77	- - anguftifolia	1	1	1	1	1
78	- - media	1	1	1	11	1
79	- - alp. Variegat.	1	1	1	1	-
80	Briza vulgaris	1	1	1	1	-
81	Cynofurus criftatus	-	-	1	-	-
82	- - cæruleus	-	1	1	-	o
83	- - paniculatus	o	1	1	1	-
84	Bromus vulgaris	1	1	1	1	-
85	- - Upfalienfis	1	1	1	1	-
86	- - tectorum	1	1	1	1	-
87	- - hordeiformis	1	1	1	1	-
88	- - peren. maxima	1	1	1	1	-
89	- - fpica Brizæ	-	1	1	1	-
90	Feftuca natans	o	1	1	1	10
91	- - margin. agrorum	1	1	1	1	-
92	- - triticea	-	-	-	-	-
93	- - rubra	-	1	-	1	-
94	- - vivipara	1	-	-	1	-
95	- - ovina	1	1	11	1	-
96	Avena pratenfis	1	1	1	1	-

Fig. 1. Title page and a page of results from *Pan Suecicus* (1749).

Although this Öland and Gotland journey, as also Linnaeus's later official travels in 1747 in Västergötland, described in *Wästgöta-Resa* (1747), and in 1749 in Skåne, described in *Skånska Resa* (1751), may have added nothing directly to the prosperity of Sweden, they contributed much to general knowledge of the kingdom, including its useful plants, and the resulting travel-books have remained popular in Sweden for their content and style. Linnaeus has been called *'Sveriges första turist'*. With equal justice he merits being considered Sweden's first economic botanist, indeed one of the world's pioneer economic botanists, important not only for his own publications but also for his influence upon his students travelling abroad with minds prepared and eyes ever open for native uses of plants.

Some Sources of Further Information

Bain, R.N. (1909). The Hats and Caps and Gustavus IV. *Cambridge Modern History* **6,** 758-784.

Drake af Hagelrum, G. (1939). *Linne's Disputationer.* Nässjö.

Hecksher, E.F. (1942). Linne's resor, den ekonomiska bakgrunden. *Svenska Linnésällskapets Årsskrift* **19,** 28-64.

Heller, J.L. (1964). The early history of binomial nomenclature. *Huntia* **1,** 33-70.

Hofmann, R. (1992). Färbepflanzen und ihre Verwendung in Österreich. *Verhandlungen der Zoologisch - Botanischen Gesellschaft in Österreich* **129,** 227-269.

Horn, E. (1893). *Die Disputationem und Promotionen der Deutschen Universitäten vornehmlich seit den 16. Jahrhunderts.* Leipzig.

Linnaeus, C. (1745). *Öländska och Gothländska Resa po Riksens Högloflige Ständers Befallning förrättad Åhr 1741.* Stockholm [Translation into English, with introduction and modern nomenclature for plants and animals, by Marie Åsberg & W.T. Stearn, 1973. *Linnaeus's Öland and Gotland Journey.* London: Academic Press. Reprinted from *Biological Journal of Linnean Society,* Vol. 5 (1973)].

Nelson, M.C. & Svanberg, I. (1987). Lichens as food, Historical perspectives on food propaganda. *Svenska Linnésällskapets Årsskrift* 1986-1987, 7-51.

Royal Swedish Academy of Sciences (Kungl. Vetenskapsakademien). (1907). *Carl von Linnés Betydelsé såsom Naturforskare och Läkare.* Uppsala.

Stearn, W.T. (1957). An introduction to the Species Plantarum and cognate botanical works of Carl Linnaeus. Prefixed to Ray Society fascimile of Linnaeus, *Species Plantarum* Vol. 1. London: Ray Society.

Stearn, W.T. (1959). The background of Linnaeus's contributions to the methods and nomenclature of systematic biology. *Systematic Zoology* **8**, 4-22.

Stearn, W.T. (1976). Carl Linnaeus and the theory and practice of horticulture. *Taxon* **25,** 21-31.